HANGING ON TO THE EDGES

Hanging on to the Edges

Essays on Science, Society, and the Academic Life

Daniel Nettle

https://www.openbookpublishers.com

ISBN Paperback: 978-1-78374-580-7
ISBN Hardback: 978-1-78374-581-4
ISBN Digital (PDF): 978-1-78374-582-1
ISBN Digital ebook (epub): 978-1-78374-583-8
ISBN Digital ebook (mobi): 978-1-78374-584-5
ISBN Digital (XML): 978-1-78374-608-8
DOI: 10.11647/OBP.0155

Cover image: Photo by Alessio Lin on Unsplash, https://unsplash.com/photos/E0LJBY360HI. Cover design: Anna Gatti

All paper used by Open Book Publishers is SFI (Sustainable Forestry Initiative), PEFC (Programme for the Endorsement of Forest Certification Schemes) and Forest Stewardship Council® (FSC® certified).

Printed in the United Kingdom, United States, and Australia by Lightning Source for Open Book Publishers (Cambridge, UK)

Contents

Introduction

> Those with any imagination and understanding are filled with doubt and indecision.
>
> –Bertrand Russell, *New Hopes for a Changing World*[1]

This is, in many respects, an anti-book. Books have a clear, unitary central message. The message is set out clearly in the opening chapter; seen growing up, fighting off rivals and doing all kinds of good deeds in a series of episodes in the middle; and then triumphantly restated at the end. Books come from certainty and self-confidence: the world is simpler than you thought! Anti-books, on the other hand, grow from critical self-reflection, compromise, and doubt. They cross and re-cross a complex landscape, trying to see its features from as many angles as possible, pointing out commonalities and false friends, abandoning one path and trying another. Their central message, if there is one at all, cannot be summarised in a sentence, but perhaps emerges, unsuspected, from an entanglement of detailed local engagements. It is a set of value commitments as much as a claim.

In 2016 I realised, with some alarm, that I had been working for over twenty years (twenty years!) at the interface between the biological and the social sciences, trying to cross the gulf that still tends to separate those two great human endeavours. What conclusions did I have from all this effort? None clear enough, right now, for a book; but plenty for an anti-book. I had been downcast for years that where other people had grand, bold theories or sweeping claims to make their names

---

1   Russell, B. (1951). *New Hopes for a Changing World* (London: Allen & Unwin, p. 5).

    https://doi.org/10.11647/OBP.0155.15

with, I did not. I had a lot of reading and thinking behind me; a lot of experimentation with different methods and ideas, without entirely nailing my colours to any of them; a lot of 'both sides have useful things to contribute' sentiments; a lot of reasonably good-humoured scepticism; and a great deal of respect for the craft. Running through all this was a diffuse sense of slight disappointment: in private moments, I could see that none of the theories espoused out there in the literature, especially those espoused by me, quite lived up to their promise. The big breakthrough had not quite come. When was I going to discover my gift?

It was only latterly that I realised: disappointment, good-humoured scepticism and the ability to see something valuable on both sides *are* gifts, of a sort. At any rate, if they're what you've got, they're what you've got. I resolved to reflect on human nature in a way that did not suggest closure, overstatement or facile answers, yet still offered something useful beyond the status quo. More than that, I wanted to find a way of writing more honestly about the academic life. The published record of books and papers airbrushes out a lot of the true nature of this life. Generally, the more influential and prestigious the publication, the more severe the airbrushing is. Readers can see only the tiny subset of thoughts and experiences that makes it through the filtering and signalling processes usually involved in publication. The quotidian mass of unpublished rumination is less cocksure, more imaginative, and in some important sense, truer. The excision of all the doubt and exploration from the final product both biases the scientific record, and gives novice scholars a completely unrealistic sense of what the academic life is really like. I have here tried to find a way of writing that is more open, more like an authentic conversation, than academic papers generally allow for. Over the course of the writing of this book, the search for the authentic voice became part of the substance as well as the style. As such, I hope the reader will forgive me an informality of tone, a periodic recourse to flippancy, and a certain self-involvement, in what follows.

Hanging on to the Edges consists of fourteen essays, written in 2016, 2017 and 2018, and originally published separately on my website.[2] My

---

2	www.danielnettle.org.uk

intention was that each essay could be read in a single sitting (ignoring the footnotes unless you are keen to follow up sources), and each would stand alone. There are, however, plenty of connections between them if you want to make them. They are organized into three groups. Part I is a set of critical journeys through the current terrain of the human sciences. Respectively, these examine the recurrent tendency for researchers to over-claim for their theories (*How my theory explains everything*); the pernicious and persistent maintenance of a false distinction between the 'social' and the 'biological' (*What we talk about when we talk about biology*); the continuum from social theories that see humans as too self-determining and independent-minded, to those that don't see them as self-determining and independent-minded enough (*The cultural and the agentic*); the perils of conceiving of culture as something like DNA and its change as something like natural selection (*What is cultural evolution like?*); and finally, the whole question of what a theory is, and what a good one would look like in the case of human behaviour (*Is it explanation yet?*).

Part II turns to the topic of poverty, particularly the consequences of poverty within affluent societies. This is one of my main specialist interests, and a topic to which the human sciences must turn with renewed vigour. Poor people know what it means to be hanging on to the edges, in many different ways. Just as the experience of poverty is insufficiently discussed, researchers struggle to conceptualise the causes and consequences of poverty in sensible ways. Thus, these essays deal as much with how to theorise about poverty as the empirical reality of what poverty is like. *The mill that grinds young people old* examines the link between poverty and ageing, and begins to raise my general claim that the behaviours of poor people typically make pretty good sense given the conditions under which they have to live. *Why inequality is bad* argues that economic inequality is bad for well-being, though not necessarily for the reason people usually say. *Let them eat cake!* considers the role of hunger in the consequences of poverty, arguing that before we turn to more abstruse and symbolic arguments, we consider the visceral: poor people in affluent societies simply can't afford to eat well, and this could explain a lot. *The worst thing about poverty is not having enough money* argues, as its title suggests, that our explanations for the behaviour of poor people should start not with their intrinsic traits, culture, or life skills, but with the elephant in the room: their scarcity of material resources. Evidential support for this argument comes from what happens when this material scarcity is lifted, even if that uplift is completely at random. Finally in part II, *Getting your head around the Universal Basic Income* examines the case for that particular form of social transfer, focussing both on its potential advantages and the intuitive reasons people find it problematic.

In part III, I return to the academic life, its institutional organization, and how one might navigate its shoals. I have always been an advocate of inter-disciplinarity, and this is the subject matter of the first two essays. In *The need for discipline*, I try, with more success than I anticipated, to force myself into the view that specialist disciplines have value and transmit useful skills. In *Waking up and going out to work in the uncanny valley*, I reflect on the difficulties of doing inter-disciplinary work for a living, relating this to the categorical proclivities of the human mind. The penultimate essay, *Staying in the game*, is a slightly battle-scarred reflection on how to survive in academia. The final essay, *Morale is high (since I gave up hope)* links the personal search for the good life to the broader difficulties swirling around science, which concern reproducibility and incentives for discovering truth.

For all I have said about this being an anti-book, there is a positive vision for the human sciences in there somewhere, albeit not spelled out in capital letters. I believe in the eventual unity of all knowledge. I believe in the capacity of science to discover the truth about the world, including about this extraordinary, self-conscious ape, and the various ways his social life on this planet might be organised. To realise its potential, science will need to be done rather better than it currently is. I look forward to a world where researchers entertain ideas without building fiefdoms; keep their ears and minds open; treat each other with courtesy and respect; seek out the places where their ideas and knowledge break down rather than comfortable confirmation of what they already believe; and admit when they don't know, or are wrong. This is also a world where we academics studiously avoid the traps of ordinary-language dichotomies and institutionalised ways of thinking; apply ourselves willingly to problems that really affect human well-being; are prepared to advocate certain social arrangements over others where we feel that our knowledge warrants doing so; and assume as much humanity and nuance in the lives of people different from us (other classes, other cultures, other times) as we do in our own. That is a world worth working for.

In closing, let me read to you from a letter written by the great Paul Feyerabend to the future readers of his final book.[3] The letter was never used in that book, but was discovered on a floppy disk (remember those?) after Feyerabend's death. He writes:

> In a few pages, you will find a story written in a style you may be familiar with. There are facts and generalizations therefrom, there are arguments and there are lots of footnotes. In other words, you will find a (perhaps not very outstanding) example of a scholarly essay. Let me therefore warn you that it is not my intention to inform, or establish some truth. What I want to do is change your attitude.

I can think of no better introduction to *Hanging on to the Edges*.

---

3   The letter is reproduced in Ian Hacking's introduction to the fourth edition of *Feyerabend's Against Method: Outline of an Anarchist Theory of Knowledge* (New York: Verso, 2010, p. xv).

PART ONE

1. How my theory explains everything: and can make you happier, healthier, and wealthier

> One has consistently to *think against oneself* — to make matters as difficult for oneself as one can.
>
> –Jon Elster[1]

The quintessence of science is doubt. In other kinds of belief systems, the statements held to be true are hardened into dogma, declared as absolute and certain. In science, they are held provisionally, constantly questioned, continually refined and replaced. At any moment, new observations or reasoning will lead to them being qualified, revised or thrown out entirely. In dogmatic belief systems, all our epistemic effort is directed towards the confirmatory instances of what we believe; see, I went to sacred waterfall yesterday, and today my cold is better. In science, by contrast, all our epistemic effort is directed to the anomalies; the cases where the prediction is not met; where the theory breaks down; the puzzling inconsistencies that help reject incorrect claims or stimulate the development of a whole new paradigm.

This is fine, and so far, uncontroversial. But I spend a lot of my time reading what scientists write, and I have to ask myself: why, then, do they seem so damned sure of themselves? Where is all this doubt that is supposed to be so great? Did these people not get the memo about the Enlightenment? It seems like doubt hardly gets a look-in in your average scientific paper or book. If it appears at all, it is mainly levelled at other people's claims. At best it plays cameo parts: for example in the opening section or two, clearing the stage of clutter for the author's grand proposal to enter. And grand it is: here is my paradigm, here is what we have shown, here are all the things my theory can do. Come into my camp; what we have in here is good. In other words, individual

---

1   Elster, J. (2007). *Explaining Social Behavior: More Nuts and Bolts for the Social Sciences* (New York: Cambridge University Press, p. 13), https://doi.org/10.1017/CBO9781107763111

https://doi.org/10.11647/OBP.0155.01

bits of scientific rhetoric look uncomfortably unlike how science is supposed to work, and awfully like how persuasion is done in non-scientific realms.

§

The overplaying of one's own correctness, and the underplaying of doubt, have a number of different flavours. In empirical papers published in the primary scientific literature, the principal flavour consists of finding a positive result, building the paper around it, and discussing it as if it definitely represented a real and important generalization about life. By 'positive' here, we usually mean some pattern that can have a statistician's p-value attached to it, preferably a rather small one. A small p-value is taken to mean that the said difference or association is unlikely to be due to chance, and hence represents an important discovery about the world. The problems with this way of reasoning are well documented. For a start, there is a good argument that 'significant' p-values do represent chance in most instances. Hang on, you say, the p-value is *defined* as the probability that the result is due to chance; so if p is small, then chance is an unlikely explanation. But here it pays to distinguish between the probability of p being small given that the finding is false, and the probability that the finding is true given that p is small. They are not the same thing.

To see why, let us say that the discovery of new and surprising things about the world is rare. By now that ought to be the case. Anyway, studies tend to be small-scale (to be low in what is known as statistical power), and so often fail to detect effects even when those effects are really there. So if you do experiment after experiment, maybe only one in a hundred correctly detects a genuine new and surprising result. Fine. But, by definition, one in twenty will produce a p-value less than 0.05, the usual cut-off for declaring a result 'significant'. So, for every hundred experiments you perform, you will get a 'significant' p-value about 5 times by chance, and about one time because you have made a genuine discovery. So, 5 out of every 6 'significant' p-values are in fact due to chance. As a very famous paper put it: most published research findings are false.[2]

I have some quibbles with the details of this argument. For a start, most published *findings* are not false. The data are not false. I think almost all scientists describe the data honestly. Rather, it is the *inference* that robust truths about the world have been discovered from these small *p*-values that is false in many cases. Also, people do not choose the experiments they perform at random, but go where pilot data and other information suggests there may be something going on. So the rate of true discovery may be higher than the argument implies. And then the argument neglects that many of the 'significant' *p*-values reported in the social and behavioural sciences are neither due to chance, nor discoveries about the world, but just reflect semantics.[3] An example is the literature on the personality trait of neuroticism as a predictor of clinical depression. This is a perfectly serious topic of investigation. However, the predictor (personality trait neuroticism) is measured with a questionnaire containing questions like 'Do you often feel blue? ', and the outcome (depression) is measured with a questionnaire containing questions like 'In the last two weeks, have you felt blue? '. It would be an odd world indeed where you didn't get an association with a 'significant' *p*-value in a study like this, because the null hypothesis (people who say they often feel blue are no more likely than anyone else to say they have felt blue in the last two weeks) is somewhat nonsensical. Its failure doesn't tell you much about humans beyond, perhaps, a basic capacity for self-consistency. It tells you more about the semantics of the questionnaires. But certainly this 'finding' is not a falsehood.

These quibbles aside, the general burden of the criticism holds. A small *p*-value does not mean that anything has been demonstrated that is repeatable. The picture is in fact rather worse than I have suggested so far. This is because of 'researcher degrees of freedom'. Basically, for any reasonably complex dataset, there are several different statistical analyses that could be performed. For example, there may be several slightly different choices for the outcome measure; several nuisance

---

2   Ioannidis, J. P. A. (2005). Why most published research findings are false. *PLoS Medicine* 2: 0696–0701, https://doi.org/10.1371/journal.pmed.0040215

3   Arnulf, J. K. et al. (2014). Predicting survey responses: How and why semantics shape survey statistics on organizational behaviour. *PLoS ONE* 9: e106361, https://doi.org/10.1371/journal.pone.0106361

variables that either could or could not be adjusted for; and several subsets of the data that could be discussed. As a result, the researcher has quite a few different goes at getting than all-important small p-value, meaning that the probability of finding one by chance is typically rather a lot higher than 0.05.[4]

An illustration of the power of research degrees of freedom comes from a recent study of large clinical trials of treatments and preventative interventions for cardiovascular disease.[5] Starting in the year 2000, the researchers carrying out these trials had to 'pre-register' their planned analyses in a public database of trial protocols. This means that they had to say, in writing for anyone to see, in advance of having looked at any data, what the critical comparisons would be, and what they would consider to constitute evidence that the intervention had a positive effect. Prior to 2000, there had been no requirement to do this; the researchers collected their data, analysed their data, and then wrote up their papers. To put the following findings into perspective, these trials are not two-bit minor research projects. These are large-scale, publicly funded, medically important evaluations done by teams of eminent clinicians and biomedical scientists.

Prior to the year 2000, 57% of the trials detected a 'significant' benefit of the intervention they were studying. After 2000, 8% did. The pre- and post-2000 trials did not obviously differ in any other respect than the requirement for pre-registration of planned analysis. It could of course be the case that human cardiovascular physiology changed in some fundamental way around the turn of the millennium, much as Virginia Woolf had claimed about human character on or around December 1910.[6] Somehow I doubt it though.

§

All in all, then, it should be no surprise that when studies in the behavioural and life sciences are replicated, we don't consistently see

---

4   Simmons, J. P., L. D. Nelson and U. Simonsohn. (2011). False-positive psychology: undisclosed flexibility in data collection and analysis allows presenting anything as significant. *Psychological Science* 22: 1359–66, https://doi.org/10.1177/0956797611417632

5   Kaplan, R. M., and V. L. Irvin. (2015). Likelihood of null effects of large NHLBI clinical trials has increased over time. *PLoS ONE* 10: 1–12, https://doi.org/10.1371/journal.pone.0132382

6   Woolf, V. (1924). *Mr Bennett and Mrs Brown* (London: The Hogarth Press).

the patterns reported in the originals; the patterns argued so confidently to be real and to be a vindication of this or that theory.[7] A great deal has been written about this recently. It makes uncomfortable reading. The scales are falling from our eyes.

As consumers of science, we have to shift our focus from thinking about individual findings as instruments of truth, and instead thinking in terms of a slowly-evolving *population* of findings.[8] If one study finds that eating broccoli is statistically associated with less depression, that's not news. Once the *population* of scientific studies starts to contain finding after finding associating, in diverse ways, broccoli with reduced depression, that starts to be something interesting; maybe even heading for knowledge counter. Even once there it will continue to be dissected and its causal basis probed. It's important for the media to understand that the development of knowledge is a gradual, population process, else it looks from the headlines like science is changing its mind all the time. This invites a kind of scepticism and hostility about the whole enterprise. Say I am trying to work out the proportion of cars on the roads that are red. Imagine if I issued a press release when I have observed a red car: 'Study finds that all cars are red!'. Then researchers at a rival university could do a debunking study. They observe a car. It's not red. 'Study finds that no cars are red. Earlier study was flawed!'. Both parties increase their scientific visibility. But the truth is: neither study is decisive (that's why science is hard); and yet every decent study very slightly increases the precision of our collective knowledge (that's why science works).

§

What interests me the most is not the near-universal over-selling that goes on in scientific papers describing the results of primary empirical studies. It is the programmatic over-selling you find when academics write their 'big idea piece', or BIP. The BIP is often a book, or more rarely, a long-form discursive article. If a book, it may be aimed at a

7   Open Science Collaboration. (2015). Estimating the reproducibility of psychological science. *Science* 349: aac4716, https://doi.org/10.1126/science.aac4716

8   McElreath, R., and P. E. Smaldino. (2015). Replication, communication, and the population dynamics of scientific discovery. *PLoS ONE* 10: e0136088, https://doi.org/10.1371/journal.pone.0136088

more general audience than just the few research specialists in the field; the so-called 'popular' or 'trade' book. It's the author's chance to cement their position in the minds of the community; to recruit future devotees; to make their mark. (By the way, the pronouns in the preceding sentence are not grammatical errors, but none other than 'singular they', the American Dialect Society's word of the year for 2015. 'Singular they' will henceforth be used throughout this book. And by the other way, let me pre-empt your inevitable thought: I am as guilty as anyone else of the sins to be described below.)

The BIP has a recurrent four-act structure.

Act One: There are huge problems to be solved. The science is in disarray. Our existing theories are bad or incomplete.

Act Two: Heroically, through alarms and excursions, I've come up with this theory. It overturns the bad theories and completes the incomplete ones. It solves the problems. Here are some arguments in its favour.

Act Three: Here is another thing my theory is good at. And another. The good things about my theory go beyond the problem I set out to solve. Within a broad domain, it is quite possibly the long-awaited theory of everything.

Act Four: You can apply my theory to life. Unification of the sciences? My theory. Economic uncertainty? My theory. Unlucky in love? You've guessed it.

We can know these acts by nick-names, each of which tells you what the problem is. They sound like the syllabus for a class on bad reasoning:

Act One: The straw man

Act Two: The great man view of history

Act Three: Confirmation bias

Act Four: If all you have is a hammer, everything looks like a nail

The BIP is, more or less, a systematic over-statement of the relative merits of the position the author is peddling. I say relative merits, because not only is the author's position inappropriately bigged up; the alternatives are inappropriately belittled too. It is not enough for my theory to succeed; others must fail. Usually, the existing theories are mis-represented to the point of denying their daily successes and the valid insights on which they were based. There is a tendency here to go for the tall poppies. Presumably you get more points for flaying a big theory than a little one. The favourite target of all seems to be the biggest: Darwin's theory of evolution. As John Welch has recently pointed out, there is a quite a cottage industry of setting out to declare Darwin's theory fundamentally inadequate/incomplete/incorrect, usually as Act One to the author's own BIP.[9]

You can see why. If you start out by saying: 'evolutionary theory is basically fine as far as it goes, but I want to talk about phenomenon X somewhere in the world of living things', then you are just a worker at the mill. You pay the rent but don't get your picture on the cover of *Nature* magazine. If you start out by saying: Darwin's theory fundamentally requires refinement/completion/replacement (by the thing I want to talk about), then you are an intellectual giant-killer, and people pay attention. But some of the claims that follow can be a little over-cooked. Often they present a phenomenon that is, directly or indirectly, the *outcome* of genetic evolution as if it challenged our fundamental understanding of the *process* of genetic evolution. You may have seen examples: Richard Dawkins says evolution is all about selfish genes, but what about mutualism/mirror neurons/mariachi music? They're really important! Evolutionary theory as we know it can't be right! It's not that mutualism/mirror neurons/mariachi music aren't interesting or important, of course. Nonetheless, I can't read this kind of BIP without the image coming to my mind of a philosopher of science, somewhere across the world, weeping quietly in order not to wake their spouse.

Whereas in Act One, all other theories are caricatured and over-criticized, in Act Two, the author's own big idea is suspiciously free of problems. It accumulates credit for things that are indeed consistent with it, but are probably consistent with many other theories too. And then we only get to visit the cases that show it to its best advantage. The odd thing here is that the author ought rationally to believe that their own

---

9   Welch, J. J. (2016). What's wrong with evolutionary biology?. *Biology & Philosophy* 32: 263–79, https://doi.org/10.1007/s10539-016-9557-8

theory, too, is likely to turn out false. Philosophers even have a name for this reasonable inference: the pessimistic meta-induction.[10] The grounds for the pessimistic meta-induction go like this: the vast majority of explanatory beliefs about the world that humans, including scientists, have entertained through our history have turned out to be wrong. So faced with a new one, a reasonable being's assumption should be that it too will turn out to be wrong. We should thus entertain a certain detachment and vigilance toward it. Yes, it could be interesting, worth thinking about, but at least in the current form, it's probably wrong and it's not going to be around for very long. The authors of BIPs are good at applying the pessimistic meta-induction to all the other theories; they just can't take the extra step and apply it to their own.

In some cases, Acts Three and Four become comical: the second half as farce. In Act Three, The Theory, extended further and further beyond any basis it had in the technical literature, becomes more and more under-specified, under-evidenced and under-grounded. It's not even a promissory note. It's a vague promise to meet you with an important package at some unspecified time in the future in a pub near Chingford. But by this point the author is three months late with their manuscript and subsisting on a diet of their own rhetoric. And then Act Four. You can almost hear the literary agent: 'Well, they like it, but it's a bit academic. It has a better chance of breaking through if readers could see how to apply it to their daily lives. Could you put in a chapter about how to make practical use of your big idea in internet dating/choosing a pension plan/promoting world peace? After all, we've got a big advance to pay off'.

This is not how science ought to work, is it? Surely the BIP should be a little bit about (a) the *common features* of different existing intellectual approaches to a problem, with a view to how they can be synthesised; and a lot about (b) the *failures* of our current understanding: the

---

10 See Doppelt, G. (2007). Reconstructing scientific realism to rebut the pessimistic meta-induction. *Philosophy of Science* 74: 96–118, https://doi.org/10.1086/520685. Note that in the philosophy of science, the pessimistic meta-induction is discussed as an argument against the realist view of science (that is, the view that science can gradually come to approximate the objective truth about the world). I am using it in a more informal and uncontentious sense: that a lot of individual scientific ideas have turned out to be wrong.

anomalies, the failed predictions, the problems, the things that don't yet fit. That's where the scientific action is.

§

The reasons why BIPs over-state their cases do not seem too difficult to understand. Do we need to point to anything more than ordinary human self-interest? People want to get their papers published. They want their grants funded. They want status. They want their book to make a stir. And a few of the best 'popular' or 'trade' books about science have made their authors surprisingly large amounts of money. So we have a perfect incentive set for ambitious writers and thinkers to over-sell their wares. What else would we expect them to do?

There's been a lot written lately about changing the incentive structure of science, for example so that the acceptance of data for publication does not depend on the size of the p-value. This obviously makes sense. It will make it harder to ignore anomalies and negative replications, and that in turn will allow more airtime for appropriate doubt. But it will not eliminate the BIP-problem (if we decide that it is a problem, a question to which I will return). BIPs are typically statements of broad theories or paradigms. In the human sciences at least, broad theories or paradigms seldom make individual 'line-in-the-sand' predictions that can be decisively judged to have failed. They make meta-predictions: I predict that it will prove more useful in approaching a topic you wish to study if you use my framework to make your predictions, than if you don't use my framework. And it's obviously hard to defeat this meta-prediction: maybe the meta-prediction is false, maybe you just didn't use the framework correctly (yet). And who knows what it would have looked like if you had used a different framework anyway? So there's plenty of wiggle-room for writers of BIPs to construct cases for their pet theories, and scope for ailing big ideas to persist.

Anyway, there may be deep reasons BIPs so often over involve over-selling. This is suggested by a recent book, actually itself a BIP, Hugo Mercier and Dan Sperber's *The Enigma of Reason*.[11] To explain their central thesis, first let us grant that there is a human capacity for reasoning, which is, roughly speaking, the capacity to produce

---

11   Mercier, H. and D. Sperber. (2017). *The Enigma of Reason* (Cambridge, MA: Harvard University Press).

and evaluate reasons for beliefs and actions. This capacity is central to the possibility of science. Mercier and Sperber review decades of psychological research showing, pretty unanimously, that human reasoning is systematically biased. Importantly, this is not just true in science, but across a broad range of contexts, everyday as well as arcane.

Specifically, human reasoning often shows evidence of 'me-sidedness'. Individuals find it easy to accumulate reasons for, and hard to find reasons against, things they are anyway disposed to believe intuitively. When it comes to things they want to believe, they will accept relatively weak reasons for doing so; indeed, when asked to generate reasons for their actions and beliefs, their initial offerings are typically weak and superficial. Only when really pushed by other people will they come up with better ones. By contrast, when evaluating beliefs or actions that other people find intuitive, but they themselves have no strong intuitions about, they take a more balanced view of the pros and cons. They evaluate other people's reasons in a much more demanding way than they generate their own.

Common experience tells us that me-sidedness in reasoning is very widespread, but it has taken some experimental deviousness to demonstrate it directly. In one ingenious set of experiments, participants selected answers to some logic problems, and gave reasons for their answers.[12] The problems were of a kind that has an undeniably correct answer, but not one so obvious that everyone sees it straight off. In a second phase, the participants were given the (different to their own) answers and reasons of another participant, to see if they accepted these and wished to change their own conclusion. Here's the deviousness, though: in one condition, the answer and reason of the 'other participant' was in fact their own answer and their own reason, whilst the one attributed to themselves was actually that of someone else. Many participants failed to detect this, because the critical switch was hidden in a number of non-switched problems. Anyway, these problems were unfamiliar, and participants were unlikely to have any settled views on them such that they immediately recognised the departure from their own position.

---

12 Trouche, E. et al. (2016). The selective laziness of reasoning. *Cognitive Science* 40: 2122–36, https://doi.org/10.1111/cogs.12303

What were the results? First, people got the problems wrong quite a lot of the time (the problems had of course been chosen to produce this outcome). Second, when invited to reason about their own initial answers, they easily generated weak and superficial reasons in favour of them. In fact, giving the participants more time to reason about their answers did not lead to them switching their initial answers very often at all, even when their initial answers were wrong. Reason, applied to the responses they knew to be their own, just tended to confirm whatever they had intuited anyway, even when it was bad. Third, and most importantly, when they re-evaluated their own answers and reasons believing them to be someone else's, they thought quite critically about them. In fact, they rejected them as invalid slightly more often than not. Reassuringly, rejection was particularly likely if their answer had in fact been wrong. Nonetheless, the force of the result stands. These were answers and reasons that *they themselves* had in fact generated about five minutes earlier. When they thought they were justifications of their own intuitions, the participants thought the arguments were fine. When they thought they were justifications of someone else's intuitions, they were appropriately and effectively sceptical.

The experiments described above uncovered me-sidedness in reasoning by making the person's own arguments appear as if they were someone else's. The converse sleight of hand—making someone else's arguments appear as if they were one's own—has also been done.[13] Me-sidedness suggests that people will like theories more just because they feel that their own intuitions have led to them, and this is exactly what the experiments showed. Participants were introduced to material about an alien planet, and to a theory about the behaviour of two species of fictional creature on this planet. They rated their degree of belief in the theory. The theory seemed initially plausible on the information given, but more and more facts were gradually uncovered until the theory started to seem unlikely. The experimental manipulation was chillingly simple: in one condition, the theory and discoveries were attributed to an 'Alex', who was presented as a researcher finding out about the planet. In the other condition, the

---

13   Gregg, A. P., N. Mahadevan and C. Sedikides. (2017). The SPOT effect: People spontaneously prefer their own theories. *The Quarterly Journal of Experimental Psychology* 70: 996–1010, https://doi.org/10.1080/17470218.2015.1099162

same theory and discoveries were attributed to 'you' (the participant). And guess what: the participants rated the theory as more likely to be true just because it was suggested that they themselves had come up with it. This remained the case once the contrary reasons started to pile in; though reassuringly, the participants in both conditions showed a decline in belief as more counter-evidence accumulated. Between them, these results seem clear, and are corroborated by many other findings. We like our own intuitions and positions, and will accept pretty weak reasons for them; whereas we are sceptical and demanding about other people's intuitions and positions.

These facts about reasoning constitute something of a puzzle. The classical view of reasoning is that its function is to help the thinker find true beliefs and adopt right actions. It's a distinctively human adaptation for making each individual better at discovering the truth about their world. On this view, the ubiquity of 'me-sidedness' looks like a daft flaw. Here we are with this great telescope, but we usually use it with a rose-tinted lens on it. Mercier and Sperber argue, though, that the original and typical function of reasoning is not for the purpose of solitary, internal truth-finding. The function of reasoning is to persuade others, in order to facilitate social interaction.

We humans coordinate our actions with other individuals to a remarkable extent, often for mutual benefit. It's hard to do that effectively given that individuals typically have different preferences, needs, expectations and experiences. These preferences, needs, and expectations are, within each individual, largely generated intuitively. We give reasons to each other as a way of bringing about smooth coordination and effective collective action. But of course, the interests of different individuals in a social group are typically only partially aligned. We want the group to coordinate effectively, but we would prefer it to coordinate effectively in doing what *we* want, not what the others want. And so reasoning is an adaptive capacity to move the intentions and plans of others towards where we already want them to be.

Viewed in this light, me-sidedness is not a design flaw of reasoning, but a design feature. Of course reasoning should be good at finding arguments in favour of our own positions—that's what it is for! Of course it should be satisfied with the minimal acceptable argument in favour of our own position—we are already persuaded of it! Our reasons only need to be good enough to get traction with others, so only if other people dispute or reject them do we need to generate better ones. But on the other side of the coin, the evaluation of others' arguments, Mercier and Sperber's position rightly suggests we ought to be quite sceptical and demanding. After all, it's really not in our vital interests to go along with other people's agendas in life docilely. Our interests are typically different to theirs, even if both parties are going to gain from coordination. We don't want to be dupes. When other people have intentions and courses of action, then, we should evaluate them quite carefully, and we should demand decent reasons why that course of action rather than another is in fact an appropriate one.

§

Applying all this to science, *of course* BIP-authors are going to over-state their cases. They are not being cynical, knowingly doing so for financial or status reasons. They are giving us the honest output of their— often impressive—reasoning processes. It's just that their reasoning processes, if Mercier and Sperber are right, are intrinsically prone to being me-sided. Thus, as a matter of course, you can expect to find all the reasons *for* the position the author lives with and few of the reasons *against*, alternative positions down-played or used as foils, and a failure to tackle difficult counter-examples. At first blush, this all seems rather depressing for science. We look to science as the paragon of objectivity, but now we end up concluding that scientists are no better, no less partisan, than politicians or quacks, and moreover that this is because of fundamental design features of human reasoning. But actually, Mercier and Sperber's thesis is not bad news for science.

For a start, what makes science revolutionary is not that the individual scientists are necessarily any better at reasoning than politicians or quacks. Why should they be; they are no more and no less human. What makes science revolutionary is the way that knowledge-evaluating processes are socialized. Individuals may often over-claim for their positions, but the scientific community has particular norms and institutions for counter-acting this: peer review of papers, critical review articles, replications, meta-analyses. In other words, the objectivity of science is not contained within the heads of the individual scientists who come up with the ideas, but rather is distributed across the community of people who review, argue, replicate, test, critique, and teach. In evolutionary biology, we are taught that genetic mutation proposes, but natural selection decides. In science, intuition plus me-sidedness proposes, but the community (eventually) decides, and it decides at least to some extent on the basis of evidence and arguments. This is why it is so disappointing when the media present 'the scientists don't agree' as a reason for dubiousness about some area of science. Of course the scientists don't agree—if they did, they wouldn't be scientists! But here's the paradox of science: by never agreeing, by always doubting, we gradually and collectively come up with beliefs we can all agree on and which we do not need to doubt.

Actually, the reasoning experiments show that people are generally quite good at evaluating arguments as long as those arguments are not their own. This means that processes like peer review and replication, when implemented and executed wisely, will tend to do some good. These processes can be frustrating and arbitrary at times. There's nothing worse, when you are just getting going with your own me-sidedness, than having to deal with someone else's me-sidedness! Nonetheless, these painful exchanges constitute a critical selective pressure that on average improves the level of correctness in the population of beliefs that the community currently holds. They drag us very slowly up a selection gradient towards knowledge—with all the usual caveats that selection is probabilistic, that selection gradients are not uniform, and that you can get stuck at local maxima in the landscape.

Even more than this, the research Mercier and Sperber review shows that people will eventually abandon and revise their own positions if their reasons for holding them are challenged in a compelling enough way. Me-sidedness means people have strong priors in favour of their own intuitions, but they are not completely immune to updating their intuitions given enough evidence and argumentation. This, Mercier and Sperber suggest, leads to a more positive view of the scientific process than that contained in Max Planck's famous claim that: 'A new scientific truth does not triumph by convincing its opponents and making them see the light, but rather because its opponents eventually die'.[14] It

---

14 Planck, M. (1950). *Scientific Autobiography and Other Papers* (New York: Philosophical library, pp. 33–4).

suggests that as long as scientists don't live in silos, as long as they talk, argue, justify their claims to each other, then individuals with a stake in the research will at least sometimes change their minds in a way that goes from worse beliefs to better ones.

§

All this casts the BIP in a slightly different light. If the community is going to decide, it needs to know what the strongest case is for each of the options it is deciding amongst. Thus, BIP authors, ridiculous though they can be, are playing a useful role in a wider drama. There is a parallel here with the adversarial legal system in countries like England. In an English court, an impartial authority, perhaps a jury, will ultimately decide one way or another. One advocate for the prosecution and one for the defence will each present the strongest possible case for their side. The advocates are expected—indeed required—to show me-sidedness for their position, to accumulate arguments for it, and to minimise the arguments against. In the proper exercise of their functions, at least one advocate will be wrong, must be wrong, and both must be partial. So too, perhaps, in science: by writing a BIP, an author makes the best possible case for the prosecution or for the defence, not because reality probably is that way, but because the best way for the impartial community to adjudicate will be to have laid out before it the strongest possible version of the case.

I can see that it is useful for the scientific community to review bold exemplars of positions that it is trying to assess. Imagine if every statement of a theoretical position were hedged around with caveats; complete in its weighing of pros and cons; exhaustive in its treatment of possible alternatives and other factors. It would be very hard to get your teeth into exactly what was at stake. As a young man ostensibly studying psychology and philosophy, I actually spent most of my time reading popular books on evolution. Why? At least partly because the evolutionists had a big idea whose universal scope and power they presented without qualification. These were ideas singing at the tops of their voices, not mumbles and apologies. Social science writing, with its frequent insistence on variegation and specificity, on multiplicities of factors, can really lose out here in the airtime of public discussion.

We seem to be in danger of completely exonerating the authors of BIPs from their absurdities. We have seemed to have freed them from even a minimal obligation of balance and good scholarship. Rather than striving to overcome their proneness to me-sidedness, BIP-authors can claim it serves the greater good of science ('just doing my job, guv'nor'). This seems to be going too far. And there is an alternative to adversarial legal systems: inquisitorial (or non-adversarial) systems. A substantial fraction of the world's population lives successfully under such arrangements. Here, the court itself is involved in the gathering of evidence, both for and against, and must come to a balanced determination. The generation of arguments and the evaluation of their merits are unified in a single office, rather than being divided across the advocates and court respectively. Should science operate more like this?

It's said that inquisitorial systems may be better at discovering the truth, whilst adversarial systems may be better at giving all individuals a hearing and hence protecting them from the inappropriate exercise of power. I don't know. I suspect science needs both modes. Really new ideas, paradigm-shifters like continental drift, are not going to get any traction without me-sided advocates. So you need the adversarial model at an early stage in the development of a paradigm. But cocky advocates strutting against one another does get a bit wearing—tribal, sterile, prone to self-congratulation and self-perpetuation. It is incapable of sorting out the details and typically does not produce synthesis. You then need wise magistrates, a lot of them.

The population needs both behaviours, and it will always get them, because some scientists take readily to the adversarial mode, while others gravitate more to the inquisitorial. It depends to some extent on one's personal balance of approach and avoidance motivations. What gets you out of bed in the morning, the possibility of glory and renown, or the terror of turning out to be wrong? For some BIP-writers, it seems to be the former; for natural magistrates, the latter (or at least, strong scruple about balance and correctness) seems to loom larger. The same personality diversity when it comes to reasoning is observable outside science too: some people want to persuade and charm their social circle, while others place a big emphasis on listening to all sides and forming a reasonable consensus.

We need to value our bold advocates, our BIP-writers. Most of them, like Icarus, will fail, but in so doing they might just open up new terrains, inject creativity, and inspire others. They need to have a decent modicum of balance and openness, though, and we need to soften their more blatant partisanship. Mercier and Sperber's thesis suggests that we can do this in science the way we do it in the rest of life—through conversation. It is through conversation that people's reasons are challenged, questioned, refined, balanced. But it has to be conversation undertaken in good faith with others whose perspectives are different from one's own, otherwise all that results is entrenchment and polarization. BIP-authors should not be rewarded for, or by, living in disciplinary or paradigmatic silos. Rather they must be engaged in friendly and quizzical conversation. We also need to make sure our institutions value and reward the quieter and wiser magistrates too. We don't currently do this enough. In promotion, funding, publication, and visibility, more thoughtful, perhaps more honest, souls often lose out. It is easy to see how this ends up happening. Indeed it is related to the broader societal pattern of more extroverted people being rewarded more in the world of work, without obviously adding more value.[15]

Anyway, I have to stop this now and work on the draft of my next book. It's called: *As far as it Goes: A Decent Theory that Isn't Revolutionizing all of Biology, and Probably Won't Change your Life*. Do you think it is going to sell?

---

15   Pehkonen, J. et al. (2010). Personality and labour market income: Evidence from longitudinal data. *Labour* 24: 201–20, https://doi.org/10.1111/j.1467-9914.2010.00477.x

2. What we talk about when we talk about biology

> Among all these pieces of information that together produce human behaviors, which are nature and which are culture? No one knows, and it does not matter in the least—in fact no one could find out, because the separation is nowhere.
>
> –Pascal Boyer[1]

The radio network NPR titled a 2013 written piece about an interview with scientist and author Adrian Raine as follows: 'Criminologist believes violent behavior is biological'.[2] Sentences like this pose a problem. The problem is that they ought to be clearly nonsense; but somehow they are not. Somehow they seem, despite all reason to the contrary, to mean something. They manage to mean something to most of the people most of the time, and perhaps even to all of the people some of the time. The same is not true in equivalent cases not involving humans. Imagine the headlines: 'Ornithologist believes bird song is biological!'; 'Microbiologist believes bacterial infection is biological!'. You take my point.

What is violent behaviour? The unwelcome violation of the body of one or more victims by one or more aggressors. The aggressors do this with their feet, or their hands—hands in fists, on weapons, or even on joysticks in remote bunkers. Sometimes hands are not needed; but here, larynxes are required, larynxes wired up to brains in a particular way. And what aggressors do changes the victim's body: her knees, her

---

1   Boyer, P. (2018). *Minds Make Societies: How Cognition Explains the World Humans Create* (New Haven, CT: Yale University Press, e-book location 4819).

2   *NPR Books*, April 30th 2013, http://www.npr.org/2013/05/01/180096559/criminologist-believes-violent-behavior-is-biological

   https://doi.org/10.11647/OBP.0155.02

kidneys, her face, or just the state of her nervous system. I am pretty sure feet, hands, larynxes, brains, knees, kidneys and faces are biological. Surely, if the Almighty did not want violent behaviour to be biological, He wouldn't have made us out of meat.

Enough, I hear you say. Of course the *implementation* of violent human acts is done using biological stuff. But what we are interested in is the *reasons* violent acts occur. And to give a useful account of the reasons we need to appeal to processes of quite a different kind to 'biology'. Compare an example: US presidential elections are implemented in some districts using paper ballots, in some using voting machines. These ballots and machines are physical objects. They, or similar devices, were necessary for the implementation of the 2016 presidential election, but they aren't an interesting part of the story of the outcome of that election (unless you think there were some pretty strange election irregularities). Explaining why the outcome was one way rather than the other requires discussion of: US demography; contemporary US social, economic and political institutions; ideologies; narratives; decisions made by individual campaigns; and so forth. Election results are delivered using physics, but there is a coherent sense in which it would be controversial and rather strange to claim their outcomes *are* physics.

In the social sciences, we find ourselves in an odd quandary regarding the explanation of human outcomes. A standard position might go something like this. Humans are biologically implemented creatures, but they have special properties. In virtue of these properties, the outcomes of their lives have *reasons* and *meanings* rather than *physical causes*; are influenced by *culture or society*, not *nature or genes*. The special properties (we can argue about what they are) have a natural, biological origin. But once the special properties are in place, they permit an infinite range of possible social histories, whose explanations are to be couched in constructs that are not, in any interesting sense, biology. They float free of their substrate. These constructs are themselves quite varied, but they include talking about (choose your favourites and pay for them at the checkout): social structures, culture, norms, institutions, discourse, individual meanings, response to incentives, agency, values, and so on.

You could defend the standard position's division of labour on the basis that social structures, meanings and agency were *ontologically* different from biology; that is, different things of a fundamentally different kind. Surely these days that would be a bit hard to justify. More plausibly, you could defend it *pragmatically*. Sure, in principle if we had complete, accurate models of how biological systems worked, then maybe social processes would start to be expressible 'biologically'. But the nematode *Caenorhabditis elegans* has only got 959 cells (302 of which are neurons), and despite decades of research we are currently unable to predict exactly what an individual *C. elegans* will do next when put on a dish. What hope, therefore, the Dutch tulip mania of 1637, or any other of the complex, historically-situated human interactions that form the subject matter of social science? In practice, we may as well stick with our familiar analyses in terms of social norms, values, supply and demand, or the madness of crowds. Either way, we end up with a division of labour in the academy where biologists and social scientists don't usually get to share the same coffee room.

§

As things stand, human biologists mostly talk about things like genes and brains and hormones, while social scientists mostly talk about a separate set of processes like preferences, culture, social capital and institutions. As a division of labour goes, it works up to a point. Both parties have come up with a lot that will stand the test of time. In the long run, though, if you have badly conceived boundaries, you are going to keep having boundary problems. Individually, any one of these boundary problems might be soluble *ad hoc*, but collectively, they accumulate and unsettle. In the end, the only way to solve them is going to be by abolishing the boundary — on the ground, and in people's minds. That is where we are with the boundary between 'biology' and 'non-biology' in the human sciences.

It's not that the standard position puts humans on one side ('non-biology') and all other kinds of creatures on the other ('biology'). That would be more straightforward in some ways, though prone to immediate and easy falsification. The problem is that the standard view puts humans *partly* in 'biology' and *partly* in 'non-biology'. For example, Tourette syndrome feels like a biological kind of thing, and I don't think 'Researcher believes Tourette syndrome is biological' would garner any headlines. In fact we don't know of any single genetic cause of Tourette syndrome, the environment appears to be very important, and the manifestations are largely behavioural. Indeed, people can suppress the symptoms of this 'biological' phenomenon through voluntary effort, to some extent (in describing Tourette syndrome, the intriguing term 'semi-voluntary actions' is used).

But violence, call violence biological, and that's worthy of a headline, though in fact the kind of individual violent acts of which Raine writes are often committed impulsively without intention. It's controversial to call violence part of biology, because you have moved the ill-defined boundary. Somewhere between semi-voluntary swearing, and impulsively getting into fights, approved 'biology' has stopped happening, and approved 'non-biology' has begun. Then you get to the 1637 tulip mania and the outcome of the 2016 US presidential election, and that's *definitely not* 'biology'.

The dual view of humans with a 'biological' part and a 'non-biological' part is not new, of course.[3] It is found in Descartes' view of humans as ordinary biological animals in their bodies, with an extra-biological soul, not shared with other animals, bolted on. It is also found in the 'restricted naturalism' of the great evolutionist A. R. Wallace. Wallace saw humans as the joint product of natural forces (evolution) and some higher power. Other animals were produced by the natural forces alone. Thus, within human experience, there were both animal bits (pain, hunger, thirst, presumably sexual attraction) and non-animal bits (spiritual, moral and aesthetic values, for example). We are hybrid beings. We lie in the gutter, but we are looking towards the stars. I contemplate the eternal, fastened to a dying animal.[4]

The hybrid view causes absolute chaos once you take it at all seriously. Which aspects of human life go into 'biology' and which into 'non-biology'? For those that end up partly in each, how do the 'non-biological' bits of the story interact causally with the 'biological' bits? The 'biology'/'non-biology' division runs down the middle of all the most important questions. Health: indubitably biological but profoundly affected by social-structural factors and policy decisions. Agriculture: a

---

3   See Benton, T. (1991). Biology and social science: Why the return of the repressed should be given a (cautious) welcome. *Sociology* 25: 1–29, https://doi.org/10.1177/0038038591025001002

4   It feels increasingly like that.

set of socially-organized processes and practices that centrally involves an ecosystem of other species. And so on.

We end up with really strange claims, like 'depression can have both social and biological causes'. (The division between 'social factors' and 'biological factors' is commonly made in medical teaching.) Surely it's a bit cumbersome to hold that the very same configuration of the brain can be arrived at for two categorically distinct, unrelated kinds of reasons. More to the point, the unity of the phenomenon, its integrated nature as the end-state of individuals with particular genetic and somatic endowments developing through particular kinds of experience in particular societal contexts, is necessarily closed to us whilst the boundary remains in place. The boundary also leads us to overlook obvious but important explanatory resources. Edmund Russell's account of why the industrial revolution happened when and where it did accords a central place to Darwinian evolutionary change; not in *Homo sapiens*, but in the cotton plant.[5] This kind of explanatory move is so heterodox from a humanities perspective that Russell has to justify it as part of a broader new 'evolutionary history' paradigm. He would not need to do this if no boundary had been in place.

When Berlin was divided in 1961, families and businesses found themselves with one part on one side, one part on the other. The boundary ran down the middle of some streets. In Berlin's railways, several ghost stations were created, where trains could pass by but not stop because the above-ground exits were in the wrong sector. At Bornholmer Strasse station, trains from *both* East and West Germany passed through, but no-one from either sector could get out. How many places are we collectively failing to explore because the standard positions of social science, and of biology, fail to provide the skills, incentives, and encouragement we need in order to do so?

§

I am as interested in the reasons the boundary continues to exist as I am in campaigning to abolish it. Many intelligent interlocutors will concede that the division into 'biological' and 'non-biological' makes no real sense when you talk to them about it in detail. But then, when

---

5   Russell, E. (2011). *Evolutionary History: Uniting History and Biology to Understand Life on Earth* (New York: Cambridge University Press).

they are tired, when they are talking to a lay audience, when they need a convenient shorthand, suddenly there it is again. There on their presentation slide, or in their written summary, or in something they say: 'Here we outline a biological explanation', they say; or 'as well as individual biological factors, social context may be important'. Put a penny in the swearing jar! All that is human is biological, and social context is a biological factor. What you have just said makes about as much sense as: 'As well as numbers, addition can involve 1, 3, 7 and 9'. Even Adrian Raine, in the interview cited at the head of this essay, says of his research: 'I've got to be careful here [....] Biology is not destiny, and it's more than biology, and there's lots of factors that we're talking about there'. So in fact, Raine reproduces the 'biology'/'non-biology' boundary; all he has done is partially moved one phenomenon—violent behaviour—a little further into biological territory, whilst endorsing the view that it is a hybrid phenomenon, subject to two categorically distinct kinds of causes.

Historians of science tend to situate the origins of the persistent 'biology'/'non-biology' dichotomy in particular influential academic ideas and positions, themselves the products of the concerns of their times. Thus, on the one hand we have nineteenth-century biologists' hard division between the immortal germ-line, to which slow, evolutionary, genetic processes happen, and the transient soma, which comes into the world from the germ line, but once there is off the leash in a short-timescale world of contingent environmental processes. It only has to report back at the end in the form of lifetime reproductive success. This hard disjunction within biological thought made our processual understanding of genetic evolution tractable under the modern synthesis of the early twentieth century, but if we don't deploy it with care, it opens up an apparent space between nature/biology (supplied by the germ line as factory standard) and nurture/non-biology (happening to the soma). In this space, dualism can fester.[6] On the social science side, we have figures like Weber and Durkheim, wanting to carve out

---

6   Fox Keller, E. (2010). *The Mirage of a Space between Nature and Nurture* (Durham, SC: Duke University Press). The dualism of germline/genes vs. soma/environment is not quite the same as that of 'biology' vs. 'non-biology'. Most obviously, for all other species, we think of both the 'genetic' and 'environmental' bits as being 'biological', whereas for our own species, exceptionally, we tend to call the somatic/environmental bit 'non-biology'.

a terrain on which legitimate and distinctive social enquiry could be conducted, as well as those in the tradition of Wallace wishing to salvage deeply-held spiritual or moral beliefs despite a growing understanding of our kinship to other species. It was in the interests of all these people to reproduce and reinforce some version of the 'biology'/'non-biology' boundary.[7]

These intellectual-history accounts are all well and good, but given the extraordinary and widespread persistence of the 'biology'/'non-biology' dichotomy (including amongst people never exposed to Weisman or Weber), I am tempted to give it an explanation that's a bit more, well, biological. Maybe the distinction between 'biology' and 'non-biology' maps onto some deep-seated way of thinking that humans are predisposed to develop and find intuitive to deploy. This would make some kind of 'biology'/'non-biology' distinction a 'cultural attractor' — that is, a cultural convention prone to emerge recurrently and persist in diverse human communities, because of regularities in the way people think, remember and communicate.[8] An appealing feature of this idea is that it would explain why: (a) at the individual level, people who have been thoroughly disabused of the 'biology'/'non-biology' distinction often reproduce it nonetheless, especially in moments of distraction or fatigue; and (b) at the cultural level, discursive traditions that initially contain no 'biology'/'non-biology' distinction often acquire one over time. I think for example of Marxism here: Marx was an enthusiastic endorser of Darwinian naturalism, and his theorising founded social relations on humans as 'active natural beings' engaging in productive interactions with the rest of the natural world. Very soon, the biological naturalism was washed out, and biological and Marxist theory seem to have rather little to do with each other, either in rhetoric or in practice, thereafter.[9]

---

7   See Benton, T. (1991). Biology and social science: Why the return of the repressed should be given a (cautious) welcome. *Sociology*, 25: 1–29, https://doi.org/10.1177/0038038591025001002; and Meloni, M. (2016). The transcendence of the social: Durkheim, Weismann, and the purification of sociology. *Frontiers in Sociology* 1: 1–13, https://doi.org/10.3389/fsoc.2016.00011

8   See the concluding chapter of Boyer, P. (2018). *Minds Make Societies: How Cognition Explains the World Humans Create* (New Haven, CT: Yale University Press) for an argument along these lines.

9   Marx in the *Economic and Philosophic Manuscripts* of 1844: 'Man is directly a natural being. As a natural being and as a living natural being he is on the one hand

§

So perhaps the 'biology'/'non-biology' distinction has been built by our culture along a natural fault line in the psychological landscape. Does that help us understand people's intuitions about where the boundary of 'biology' lies in human affairs? Does it, in short, help us understand what people are talking about when they *don't* want to talk about biology?

In life, people are understandably concerned to distinguish processes that could not, through any sequence of our actions, come out any differently, from those processes where it matters what we decide to do. For example, it doesn't matter whether or not I try to persuade people that human hearts should be on the left side of the body. I don't need to bother. Nearly all human hearts are going to be there, for a long time into the future, regardless of what I do. On the other hand, the level of social inequality in Britain is related to specific actions people decided to perform at particular times. It is related to these actions in a complex way; the actions are many, the consequences are subtle and at times unforeseen; the people performed them under exposure to particular discourses encouraging them to think in particular ways. But nonetheless, I could take actions that might have some effect, somewhere down the line, on the level of social inequality in Britain.

Thus, it feels like there is useful intuitive distinction between the stuff that you just have to accept, and the stuff that could come out differently (it was different at other times, it is different in other places, or it could be different if we organized things differently). You can see how this fixed/non-fixed divide could be useful to think with, in all kinds of human contexts. Which aspects of my potential spouse do I need to just put up with (her height, for example), and which ones might I manage to negotiate or shape so they are different in the future (her behaviour,

---

endowed with natural powers, vital powers — he is an active natural being' (p. 69). Download at: https://www.marxists.org/archive/marx/works/download/pdf/Economic-Philosophic-Manuscripts-1844.pdf. Gramsci in the Prison Notebooks (written between 1929 and 1935): 'Philosophy cannot be reduced to a naturalistic "anthropology": the nature of the human species is not given by the "biological" nature of man' (Hoare, Q. and G. M. Smith eds., 1971, Selections from the Prison Notebooks, London: Lawrence and Wishart, p. 335). Of course there is a great deal more to be said about the difference between Marx and Gramsci, or indeed early and late Marx, than this glib observation.

maybe)? Reasons play a different role in the fixed and non-fixed cases. It's not important for human societies to reason well about why hearts are on the left. They might figure out a reason why they are on the left, and that would be interesting. But it's not important to the outcome how people reason about it. Not so social institutions, laws, taxation and so forth: here, the quality of the reasons we come up with affects the social outcomes we get. There is also a connection (though not a completely simple one) between non-fixity and voluntariness or intention: that which is not fixed can perhaps be voluntarily or intentionally addressed. Voluntariness and intention have special roles in human justifications for action, and hence moral culpability (you usually don't blame me for that which I did not voluntarily choose to do).[10]

The spatial metaphor of inside and outside, or body core versus body surface, often gets fused to the concern with fixity and non-fixity. So we say things like 'Deep down, he's always going to be selfish'; 'is this really *in* her nature, or is it just something *on* the surface?'. This spatial translation of the fixed to the middle and the malleable to the edges recalls some diagram of essence and accidents from Medieval philosophy. It is very intuitive, even if it makes no literal scientific sense.

§

It seems, then, that we have plenty of intuitive raw material for the cultural emergence of a 'biology'/'non-biology' distinction. We take what seems to us fixed and stick that in a category. We call this bit 'biology' to the exclusion of the rest. This is supported and made more compelling by the intuitive relation of the fixed/non-fixed distinction with the inside/surface metaphor: the fixed/'biological' is the stuff on the inside. Inside a body you find muscle and blood and viscera (and if you look closely enough, genes), stuff that you don't know how to change, that looks just like the insides of other animals. On the surface of bodies you find all kinds of things you can take on and off like clothes and ear-rings and smartphones, and these don't look 'biological' at all.

---

10 The connection between intention and moral culpability is apparently universal, though its strength may vary somewhat across societies: Barrett, H. C. et al. (2016). Small-scale societies exhibit fundamental variation in the role of intentions in moral judgment. *Proceedings of the National Academy of Sciences* 113: 4688–93, https://doi.org/10.1073/pnas.1522070113

The voluntary/involuntary distinction often gets mixed in here, too. I didn't do it because of my genes, I did it because I wanted to! Or: he didn't do it intentionally, he was compelled by an inner urge.

With this rather muddled set of distinctions culturally available, 'biology' becomes the category for everything we don't want our destinies and our social arrangements to be. 'Biology' becomes the place where all the fixed stuff lives, so if you are interested in change or contingency, you define what you do in opposition to the 'biological'. This is why, in social science, the bad word that follows the bad word 'biological' is usually 'determinism'. 'Biology' is the bit of you that is a mere zombie, compelled to follow preordained urgings. You would naturally want that bit to be as small as possible if you valued your autonomy. Finally, 'biology' is somehow inside you, with all the disgusting smelly stuff you only look at in medical contexts, whilst most of things you enjoy in life involve your outer surfaces exchanging energy with the world around you. This may explain the tremendous media air-time you can get with neuro-imaging studies showing with pretty maps how activity in the brain differs between people who are and are not X, where X is suffering from schizophrenia, falling in love, growing up in poverty, or listening to Mendelssohn. 'Gosh!' we say, 'I didn't realise that was actually happening *on the inside*'. Of course those activities involve being different on the inside. How could they possibly not? But showing that something is going on inside in no way constrains the importance of stuff going on outside in influencing why those experiences happen as they do.

§

Once the two receptacles of 'biology' and 'non-biology' have been made, incoming traffic gets diverted into either one or the other. Genes — 'biology'; environment — 'non-biology'; innate — 'biology'; learned — 'non-biology'; evolved — 'biology'; acquired — 'non-biology'; nature — 'biology'; culture — 'non-biology'. Maybe that's not such a bad thing, you say, at least as a first approximation. I've argued that fixed/non-fixed and voluntary/involuntary are actually useful distinctions to make in everyday life—that's why they exist. So maybe the conventional 'biology'/'non-biology' description is heuristically useful in scientific discourse too, at least as a rough framework for starting out? If it didn't exist under the labels 'biology' and 'non-biology', we would have to invent it anyway using different terms.

This is a reasonable view, but, I think, wrong. The 'biology'/'non-biology' distinction of the standard position has only bad features that I can see. It cuts natural continua, such as fixity, into artificial dichotomies, leading to pointless and unproductive contestation about boundary cases. Everything *could* be different; the question is more 'what would it take to make X different? '. It puts into the same category concepts that are in fact quite distinct. 'Genetic' is not the same as 'evolved'. And finally it puts into different categories things that are not exclusive. Learning, for example, is something done by genes. Yes, genes, the product of a history of natural selection, do not merely enable learning by their presence. They are intimately involved, through their expression, in how learning actually works. So when you appeal to learning, you are appealing to a genetic process (and, indirectly, to evolution). More generally, any attempt to mark the 'inside stuff' off from the 'outside stuff' is a dual disservice. It ignores how profoundly environmental processes become embedded in the body, and how things like genes can exert effects outside the body envelope.[11]

§

My thesis, then, holds out both good and bad news for the standard position demarcating social science. The bad news is: it's all biology. Everything social scientists do is biology. It's not that it will be replaced by biology in the future, in some *Who's Afraid of Virginia Woolf?* nightmare scenario. Everything social scientists do already is, and always has been, biology. Why? Because biology is the study of living things. Humans are living things, and so whatever they do, however they organize themselves, whatever extraordinary technologies they create, whatever meanings they entertain, reasons they give or tastes they develop, these are all biological processes.

All you have done here, you might respond, is to *define* biology so as to include all of the things the social sciences are interested in. We do not immediately understand social phenomena any better by doing this. Has anything actually been gained by this move? I might retort:

---

11   Dawkins, R. (1982). *The Extended Phenotype* (Oxford: Oxford University Press).

well, it was the standard position that started this, by defining biology in a restricted and artificial way so that the things social scientists are interested in would *not* be in it. All I am doing is restoring a more sensible baseline. The redefinition doesn't change anything overnight, but in the long run it might help us do better research. As scientists, we are *bricoleurs* (tinkerers), using whatever materials and ideas we have lying around to try to solve new problems, or better solve existing ones. If, by erasing boundaries within the academy, we expand the repertoire of techniques and ideas that individuals know and encounter, then we might all be able to progress a bit more quickly.

So the bad news is: it's all biology. But here's the good news: biology is not what you think it is. The resistance to 'biology' in the social sciences is founded on the fear that certain things we really value would be foreclosed by joining up to the biological sciences. I believe this fear is largely groundless.

The fear arises from an opposition between law-like determinism (assumed to characterise biology as a natural science) and the world of partial influences and historical contingencies (assumed to characterise the social sciences). In natural science, the argument goes, there are absolute laws. If you drop a cannon-ball from a tower on earth, there is no doubt about the outcome. In the social sciences, we have general tendencies and patterns, but these always have exceptions and specificities in their realization; we have historical processes that are explicable in retrospect but could not have been predicted prospectively. We social scientists just can't fit what we do into a world of simple natural laws.

This fear is easily dismissed, since its view of how natural science works is misleading. That science consists in uncovering a few simple and absolute laws holds, if it holds at all, only as a description of physics. Philosophers of biology are pretty clear that biological science is not like this. The history of life is a contingent and path-dependent historical process. There are certainly regularities in the way it has evolved, but its course is a complex resultant of selection gradients, available variation, the kinds of raw material that was already there, chance, and time. What evolution produces is an astonishing diversity of inter-linked *systems*: cellular systems, organ systems, organisms, social groups, ecosystems. These are all dynamic; their dynamics depend on where they start from, and feed back into the selection pressures of the future. Although sense can be made of how these different kinds of systems work, no-one believes you can simply read off all their specific properties from a few very simple laws, either of an evolutionary or biophysical kind.

A second fear is that biological science involves a kind of explanatory monoculture, whereas explanation in the social sciences needs to be more heterogeneous. The social sciences involve identification of many different mechanisms of rather different types, at different levels: individual-level mechanisms like response to incentives or psychological biases; social-level mechanisms like stratification or spatial assortment; cultural-level mechanisms like diffusion of innovations; or even symbolic and discursive-level mechanisms. Critics of social science see the diversity of this explanatory menagerie as a weakness. A field with such an undigested diversity of explanatory strategies must just be a conceptual mess. But a lot of social scientists would respond that they value this very diversity. In complex human social phenomena, you can't just deduce the historical outcome from properties of the individual psychologies, or the social organization, or cultural diffusion, alone. You need all of these things, and whilst you need to understand how they inter-relate, you can't eliminate any of them from the stock of things we need to appeal to, and should not try.

The fear of 'biology' that comes from this somehow assumes that biological science, in contrast, only admits of one type of explanatory construct. That explanatory construct, in this straw biological science, is usually molecular (for example, a gene, a hormone, or a neurotransmitter). This is why another bad word that often follows the bad word 'biological' in social-science-speak is 'reductionism'. Once again this is a gross mischaracterization. Biological science is a diverse enterprise involving people who work at many levels. Almost all of the levels involve systems thinking: from cellular systems, physiological systems, whole organisms, swarms, hives, communities, populations to ecosystems. Though researchers are centrally concerned with how the functioning of the systems at one level relate to the dynamics at another level (e.g. molecules to cells, cells to whole organisms, individual organisms to populations), the traffic goes both ways, and there is certainly no simple theoretical monoculture. The stuff that goes on inside the individual is not, in principle, theoretically privileged over population processes. There is no sane proposal to eliminate the organism, the population, or the ecosystem as levels of analysis, or to deny that there are complex dynamics at these levels we cannot simply deduce from the dynamics at more molecular levels. In fact, explanation in biological science looks somewhat like explanation in social science: it is not a matter of reducing everything to the molecular level, so much as the identification of various kinds of mechanisms, operating at different scales, in slightly different ways in different contexts, to shape the outcome of complex and variable systems through time.

The cause has not been helped here by the writings of certain zealous 'biologisers' from the humanities and social sciences. Look at the social sciences, they say, a complete hodge-podge of unintegrated, theoretically incoherent sub-disciplines, not really getting anywhere. Now look at biological science. It's the most successful branch of knowledge of the past hundred years. It is conceptually unified by the theory of evolution. If the theory of evolution did that for biological science, then given that we too are evolved beings, it can now do the same for the social sciences and humanities. I have sympathy with many aspects of this view, but it is important not to over-simplify for rhetorical effect. Inspired by popular accounts of evolutionary biology, the 'biologisers' ascribe biological thought all manner of positive properties they feel the social sciences don't currently have, like simplicity, unity, and theoretical elegance. A lot of these properties turn out to be somewhat over-stated once you start actually swimming in biological waters. There is widespread human tendency to under-estimate the complexity and internal heterogeneity of categories of which we don't have much direct personal experience. I suspect there is an inverse correlation between how unified and elegant you think biological science is compared to social science, and how many hours you have ever spent in a biology lab or field site.

The truth is that biological science, viewed from closer up, is also something of a hodge-podge; less so than social science, but a little bit hodgy and in some respects podgy nonetheless. It is unified by the theory of evolution only to a point. Most working biological research is cellular and molecular, and here the theory of evolution usually plays rather little role either in techniques, explanations, or the kinds of questions people ask (it probably should do more, but it doesn't at present). And then even in the parts of biological science that are more explicitly evolutionary, such as whole-organism biology and ecology, there's a great deal we don't know about how it all works out in detail. The 'biologisers' sometimes imply that it suffices to read Fisher's *The Genetical Theory of Natural Selection*, maybe coupled with William Hamilton's seminal 1964 papers on kin selection, and then you can simply deduce kidneys, or star-nosed moles, or what happens in Yellowstone National Park. You can't. That's what biologists do all day. They don't agree about it all. It is going to end up involving a great diversity of processes, and a lot of detailed understanding of specific mechanisms.

A final fear that social scientists might have about changing the sign over the office door to 'biologist' is that they would lose the opportunity to speak about agency, and related notions of responsibility and the moral life.[12] The non-human living world may exhibit complexity, but it is not a world whose inhabitants have agency. This is why human affairs are different, and we need to hold on to this fact in the ways we talk about it. I agree that agency, and the things that go with it, are distinctive features of human life that need to be accounted for, not down-played or ignored. But 'distinctive' need not mean 'not biological'. In fact, some of our existing social science theoretical frameworks do a bad job with agency. Social constructionism would be one example. If our very personhood is a cultural construct, originating in social discourses we were exposed to but did not choose, then what sense can we make of responsibility, moral justification or voluntary action? Are they not just shams? The more I think about agency, the more I feel that it is not so much that agency *can* be reconciled with seeing us as biological beings. It is that *only* by seeing us as biological beings can we rescue any coherent notion of what human agency is or the uses to which it is generally put.[13] Joining up with biological science certainly does not make the problem of agency any more difficult than it already is. If anything, it's the key to progress.

§

---

12   For some discussion, see Scruton, R. (2017). *On Human Nature* (Princeton: Princeton University Press).

13   See *The cultural and the agentic*, this volume.

Re-designating social science as one of the biological sciences should not be a case of *restricting* the types of data or explanatory entities to which social science is entitled. And it is certainly not a case of seeing humans as *just the same* as monkeys or mongooses. After all, monkeys are not *just the same* as mole rats or slime moulds, yet they are all totally biological. It's more a case of feeling free to pursue the theoretical and empirical connections between different types of data, types of description, types of process, without getting held up at the border post. It gives us a greater library of options to improvise with, connections to make.

The goal of abolishing the 'biology'/'non-biology' distinction is worthy. Can we succeed? I don't know. There are institutional and organisational issues here that are going to be slow to unpick, if they can be unpicked at all. And if the 'biology'/'non-biology' distinction really is a cultural attractor, we are running into quite a strong psychological headwind. But folk psychology is what science is here to rise above, not something to which it is condemned. Admittedly, though, people have been trying quite seriously to dismantle the boundary between the 'biological' and the 'social' for at least fifty years, and yet there is still plenty of evidence of it on the ground.

Taken together, the folk psychology of the audience, and the ways institutions are divided up, can provide incentives for perpetuating the boundaries even amongst those who know much better. In my experience, when a scholar defines their position as *not-just-X*, where X is, for example, 'biology', they are not usually interested in the actual contents of category X. In fact, they usually present a deliberately impoverished view of what those contents are. They almost *require* such a limited view. If they admitted too much of the truth about X, such as its internal heterogeneity and potential for future change, their appeal to *not-being-just-X* would probably fail in its functions. The appeal is a territorial claim; a rallying point; a stoking of prejudice; a parochial code for fellow-feelers to identify one another; it is a kind of aesthetic, moral and financial self-justification; it is a signal of social distinction. The unfortunate thing is that scholars, like politicians, can get rewarded for these kinds of moves, and misunderstood or ignored if they fail to make them.

We can try to do something about this, even if it is an uphill battle. As authors, we can catch ourselves every time we lazily or parochially use a phrase like 'Whereas biologists believe...', or 'Whilst economists see....'. Which biologists? Which economists? And why label them by their kind? Why not just use their names? As readers, we should be as critical and sceptical of parochial claims from our scholars as from our politicians. We should see ourselves as citizens of the whole wide intellectual world, and demand reasons, expressed in clear language, that make sense universally, without appeal to the tribal affiliations of their originators.

3. The cultural and the agentic

> Culture is thus an *effect* as much as a *cause*...
>
> –Herb Gintis[1]

The late Pat Bateson used to tell a joke about two philosophers. 'Thinkers can be divided into two kinds', says the first philosopher, 'those who propose dichotomies, and those who reject them'. 'Nonsense!', replies the second.

At risk of similar tendentiousness, I contend that there are two major styles of social explanation. They have their origins in lay talk about reasons for actions, and they run through the professional discourses of social researchers. Extremes of both styles fail, and they fail in complementary ways. Our job is to make synthetic theories that capture the valid insights of both styles of thinking whilst also transcending them. I will call the two styles *cultural* and *agentic* respectively. I am not sure these are perfect names. There are various synonyms and near-synonyms for each one knocking about. Nonetheless, my chosen names are reasonably memorable and I will stick with them here.

First, let's point out that human societies have an order, an order that transcends the minute-to-minute decisions, or even the lifetime decisions, of any one individual or interacting dyad. When I wander down the lane to the bakery, I don't have to devise a strategy for making it clear to the baker that the object of my desire is the leavened wheat product on the shelf. There's a convention that both of us unthinkingly subscribe to but neither of us invented, of denoting this with the sounds 'loaf of bread'. I don't have to offer to write the baker a scientific paper, teach her programming, or tend her garden (about all I could offer in improvised barter) in exchange for her comestibles. We have an institution called money by which exchange of anything for anything is

---

1   Gintis, H. (2017). *Individuality and Entanglement: The Moral and Material Bases of Social Life* (Princeton: Princeton University Press, p. 153).

 https://doi.org/10.11647/OBP.0155.03

possible and requires no further contestation other than specification of the price. I also understand that I am not to haggle over the price; that she will sell me her bread at the same price regardless of how many loaves she has left; and that it would be completely unacceptable for her to ask different prices of different customers, for example on the basis of the colour of their skins. These last two rules (price unrelated to scarcity, price unrelated to identity of buyer) are particularly interesting. Strategic agents making improvised decisions might well, you'd think, charge more when supply seems short relative to demand, or when selling to people they don't like, but in the shops on my street, there are social rules that you don't do that. And these rules have moral force; people treat them as if they were binding and are outraged when they are violated. So there is a lot of social order going on, even in the simple act of procuring my lunch. This order—its structure, diversity, and evolution—is to social theory what fritillaries and swallowtails are to lepidoptery.

You are informally offering what I call a *cultural* explanation every time someone asks you why we don't haggle over price in Newcastle, or voluntarily eat horses, or allow polygamy, and you answer: that's just our culture. That's what is normative here. For this to constitute anything like an explanation rather than just a restatement of the phenomenon, you must be claiming something along the following lines. The social order itself, or something that encodes it, has a real concrete existence external to individual actions, is causally primal in respect of those actions, and hence explains those actions in a non-trivial sense. Individual actors *inherit* and *reproduce* this order, with not much more deliberation and choice than when we inherit and reproduce our DNA. Whereas the inheritance and reproduction of DNA happen by meiosis and mitosis, the inheritance and reproduction of the social order happen by socialization. To the extent to which people seem to be exerting free choice, they are only doing so within the constraints and set of acceptable roles that the social order makes available to them, like alternative expression levels of the same DNA sequence. If we want an explanation for the existence of the social order, we need to move to a different level of analysis, in which the explanatory forces will be something other than individual choices, since individual choices are the consequence, not the cause, of the social order under which they occur. Thus, in cultural thinking as I define it here, the social order, or the cultural rules that encode it, is the upstream source of individual actions.

Different flavours of cultural theory abound in social science. According to certain versions of the 'cultural evolution' paradigm, humans have a very general propensity to acquire and internalize whatever is normative in their culture. They automatically adopt the norms of the majority of people they encounter, or of the most locally-prestigious people they encounter. They do so, according to the theory, largely *credulously*; that is, without regard to how those norms suit their interests. So much so that, according to one hypothesis, whole societies can and do fail through slavish adherence by their members to self-injurious norms, in what has been termed cultural group selection.[2] The reason that the societies we observe have fairly sensible norms is not that the people in them exerted good sense, but that all the ones that happened to have bad norms have gone extinct. That's a pretty strong claim.

To take another example, in cultural theory in the tradition of Michel Foucault, the very ideas that people can conceive of as true, reason about, or discuss, are the products of a symbolic order that pre-exists them, known as an *episteme*.[3] The hegemonic *episteme* of the age controls what seems right and natural, and limits people's understanding of the current world or possible alternatives to it. You might think you hold beliefs because they are true, or do things because you want to, but really the *episteme* has constructed you to think that way. It is

---

2   See for example Henrich, J. (2004). Cultural group selection, coevolutionary processes and large-scale cooperation. *Journal of Economic Behavior and Organization* 53: 3–35, https://doi.org/10.1016/s0167-2681(03)00103-3; Richerson, P. et al. (2016). Cultural group selection plays an essential role in explaining human cooperation: A sketch of the evidence. Behavioral and Brain Sciences 39: e30, https://doi.org/10.1017/s0140525x1400106x. In fairness to these authors, their cultural evolution framework is not incompatible with individuals being somewhat discriminating about the norms they adopt, or having individual agentic preferences in addition to a general tendency to conform to norms. The models and interpretations they present do however strongly stress docile conformity, hence culturality in my sense, at the expense of purposive agency. I should also point out that what I describe here as cultural group selection is only the first of three distinct processes described under that name by Richerson and colleagues.

3   Foucault, M. (1970). *The Order of Things: An Archaeology of the Human Sciences* (New York: Pantheon).

insidiously guiding and constraining you at every turn. It is a causal force. And where does the episteme come from? Again, its origins lie at some different level of analysis, upstream from everyday behaviour and apparently voluntary individual choice.

So these are examples of how the cultural corner looks. What about the agentic? Here the leading professional representative is the easiest to characterise: the 'rational actor' model of microeconomics and political theory. In agentic theories such as this, the individual is a sovereign decision-maker. She has a set of preferences over different bundles of outcomes. These preferences are causal and primary, though of course they can vary according to the context—effectively, the set of options available—and the state of the chooser. The social order, rather than being causal, is the downstream consequence of the preferences of many agents interacting over time. If I prefer Strauss to Stravinsky, it's not that the sinister episteme has socialized me with normative conceptions of what music can be. I just like it. If Strauss is popular, it's because many other agents share my preference. If the divorce rate goes down, it's not that the social system 'needs' more stable families, or subtly coerces people into particular matrimonial roles, or even that a particularly virulent cultural meme has taken hold. It's simply that under the current set of economic and demographic circumstances, more people are finding that staying married is an attractive strategic option relative to the option of leaving their marriage.

How could an agentic theory deal with the existence of the social order? Social interactions occur, according to agentic thinking, when both parties prefer having them over not having them, and both parties seek the form of interaction that comes highest in their register of preferences. The type specimen is the mutually-beneficial exchange between the buyer and seller of a material good. You do it when it suits you; if it didn't suit you, it wouldn't happen. But the rational actor analysis does not need to stop at monetized exchanges between strangers: a scientific collaboration, a romantic relationship, or a commitment to a voluntary organization can all be captured by the same logic. The interactions go on as long as they provide something for which both parties have a preference relative to the available options for the other possible uses of their time and energy.

Taken too literally, rational actor thinking might seem to imply that there are no norms at all, just atomic strategic agents. That is, every time I go into the bakery, the whole business of fixed and universal prices is up for potential renegotiation; or every year in a marriage, each partner has to check their current preference function and negotiate the terms of sexual exclusivity for the next accounting period. But this claim would be unfair on the rational actor model, and since being unfair to the rational actor model is something of a convention in much of social science, it is important to get this right. The rational actor model need have no problem with the idea that there are norms, or even that the norms are in some sense binding. The point is that the norms themselves should be analysed, at a deeper level, as the outcomes of the preferences of interacting individual strategic agents through time. So the social order is the *dependent* variable, with human preferences as the *independent* variable, whereas in cultural explanation it was the other way around. We can analyse the norm of fixed and universal prices in the bakery as *if* it were the product of a voluntary agreement between customers like myself and the baker. In a society where good bread is abundant and affordable, I would prefer not to have to enter a great long discussion about how much money I should hand over every time I want a loaf. The baker for her part values my repeated custom more highly than the opportunity to make an extra pound or two on a one-off occasion when I am desperate, and knows she will gain this by offering me the convenience of fixed prices. So in equilibrium, it's always £3, and both parties accept the norm.

There's always a danger that agentic thinking will become Panglossian: whatever social order exists is necessarily for the best, since free agents have brought it about by acting in accordance with their preferences. This is an interesting contrast to cultural thinking, which is very often dystopian and critical of the current order. The impetus behind much cultural theory, for example, is the drive to expose the subtle roots of domination and oppression inherent in the *episteme*. But we can remain agentic without being Panglossian. The social orders that come down to us, even though they are expressions of past agents' choices, may not be optimal for the way we live now. People have differed historically in their power and control of resources, and hence social orders have been produced that favour some people at the expense of others. It is then rational to try to change them to make them fairer. Plus, importantly, there are many scenarios where benign equilibria cannot spontaneously be reached without bringing other kinds of institutions into being. For example, in 'tragedy of the commons' type situations, individuals acting in an uncoordinated way will end up at an equilibrium that none of them prefers. The solution is to advocate a higher-level mechanism of enforcement, such as a system of fines, laws or customary rules. This higher-level binding social mechanism requires active, deliberate work to bring it about; but it is still in some sense the outcome of the preferences of rational sovereign agents.

§

What are the strengths and weakness of cultural and agentic thinking, respectively? They are the mirror images of one another. The cultural view correctly captures the insight that the social order is *cumulative*, *historical*, and has *emergent properties*; and it does so more naturally than agentic thinking does. The English political system, for example, is of such a Byzantine complexity that no two freely-interacting representative agents could possibly come up with it in a reasonable time. Its peculiar design features—a second legislative chamber that consists partly but not mostly of hereditary aristocrats, the fact that the monarch is the head of state and yet wields no power, and so forth— probably don't instantiate the preference functions of any of British citizen, living or dead. They represent the current snapshot in a kind of descent with modification process. This process has a historical quasi-life of its own, so much so that the cultural perspective is in some ways right to see the institutional system as the unit of analysis, and individual politicians as partly its current vehicles, rather than the other way around. Many of the properties of the system, although they may have arisen from the voluntary decisions of certain individual actors, were not predictable from those voluntary decisions, and certainly do not represent the outcome of the actors' intentions. When non-independent, non-omniscient, socialized agents interact through time, this generates emergent properties and historical continuities. And if this applies to the English political system, it applies too to any system of meaning, knowledge, social organization or technology in any human society.

The other great strength of the cultural perspective is that it captures the fact that the social order is, for each of us and to at least some extent, *unchosen and empirically real.*[4] People do often adhere to social norms, even when there is no prospect that a violation will be detected or punished, just because they are the norms. Behaving normatively thus cannot be completely reduced to some immediate strategic or prudential calculation, such as the desire to avoid getting a bad reputation, although such concerns do of course exist. And the available ways of talking about a topic do undoubtedly influence the possible actions we entertain: this is true in science as much as in political life. These ways of talking are, in a broad sense, inherited from our culture, and often go unquestioned much of the time.

But the weakness of the cultural way of thinking is in its inability satisfactorily to account for social change; and hence, in the end, to explain which properties of the social order endure. People follow the norms they are socialized into, except when they don't. Sometimes they say: I don't want to do that any more! It's lame! It's not right! I protest! It may be what my parents did, what most people do, what the prestigious people do, but I don't feel it suits my interests and I will abandon it/modify it/flout it. So in an important sense, people are faithful replicators of norms *only when* those norms suit their current perceived interests and opportunities to a reasonable extent; otherwise, they try to change them in decidedly non-random ways. If you don't allow for this in your social theory, allow for the order-transforming, purposive exercise of human agency, you really have no useful account of how societies develop, or how we end up with the historical conflicts and compromises that we do.

Another interesting difference between cultural and agentic thinking concerns the extent to which social groups are conceived of as *heterogeneous*. Cultural thinking leads us to think of each society

---

4   Or at least, the social order feels like an empirically real object that can be appealed to. As Pascal Boyer has recently argued, our widespread propensity to refer to and reify things like 'English culture' or 'the Dinka social system' is really just a folk-sociological shortcut, something humans impose on the ceaseless and variegated flow of social interactions as a mental simplification—a simplification that can become badly misleading if we take it too literally in scientific theorising (see Boyer, P. (2018). *Minds Make Societies: How Cognition Explains the World Humans Create* (New Haven, CT: Yale University Press).

as having a perfectly shared norm structure, ethos, or *episteme*.[5] This might well be different from that of a different society or historical period, but, within the society or historical period, everyone subscribes to it; within-group homogeneity and between-group variation. Hence the anthropologist's fantasy that you can talk about 'the culture' or 'the norms' of the Fuegians or the Russians as more than a statistical summary of a distribution. But this really is a myth: careful study actually reveals that there is, in many domains, much more variation in values and behaviour within cultural groups than there is between them.[6]

For agentic thinkers, by contrast, the central and most important truth about human societies is that they are made up of diverse individuals with *different* ideas, preferences and interests. The social order is partial and contested; to the extent that it exists, it is the uneasy truce in innumerable arguments and compromises between people with different wants and values. It is always, therefore, unstable and provisional, imperfectly agreed, and will be obeyed unevenly. At every moment it will be challenged, disputed, violated, transformed and renegotiated, usually in small ways and occasionally in larger ones. It is this ceaseless seething of variation and challenge that gives the system its dynamism, but also which makes it alight over historical time on consensual, or at least hard-to-overturn, solutions to the problems of communal living.

Culturally-oriented writers do sometimes talk about within-group heterogeneity in values, and of individuals challenging the social order. For example, people may contest or resist the dominant *episteme* rather than just accepting it, according to their own context and perspective.

---

5   For example, Foucault: 'In any given culture and at any given moment, there is always only one 'episteme' that defines the conditions of possibility of all knowledge, whether expressed in theory or silently invested in a practice.' Foucault, M. (1970). *The Order of Things: An Archaeology of the Human Sciences* (New York: Pantheon, p. 168).

6   Bell, A.V, P. J. Richerson and R. McElreath. (2009). Culture rather than genes provides greater scope for the evolution of large-scale human prosociality. *Proceedings of the National Academy of Sciences of the United States of America* 106: 17671–74, https://doi. org/10.1073/pnas.0903232106. The emphasis of these authors is rather different: they show that there is more within-group homogeneity and between-group variation in cultural values than in genes. True, but let's put it into context: the rather little of the variation in cultural values that is explained by cultural group membership is a bit more than the virtually none of the variation in genes that is so explained.

This immediately raises the question: what are these people standing on whilst they try to rip up the floor? Where is the vantage point outside of an *episteme* from which you can critique it and formulate a different one more suited to your interests? What resistance to dominant ideologies tells us is that it can't be socialization all the way down; the cultural explanation, to be complete, requires that there is an agentic bedrock people can stand on to sometimes contest the *episteme*. Malleable as people may be to the norms and discourses of their time, there are some *properties of natural agency*—some no-doubt fuzzy set of recurring intuitions, motivations, judgements, and interests—that stand beyond the local social order, and which people can use to resist, criticise or reinforce that order. Maybe these properties of natural agency are only intermittently visible, as it were distant mountains often lost in the epistemic fog, but they must be there somewhere. Their roots, it is reasonable to suggest, lie in the legacy of the deep natural history shared by all living humans.[7]

§

As things stand, the social sciences are balkanized between agentic parts and cultural parts. Economics is largely agentic in its style of reasoning, and much of political science follows suit. My own disciplinary watering hole, behavioural ecology, also favours the agentic.[8] By contrast, parts

---

7 Shared by all humans except, it seems, for Michel Foucault. After Noam Chomsky, a great believer in humanity's shared properties of natural agency, debated Foucault on television in 1971, he commented: 'I liked him personally, it's just that I couldn't make sense of him. It's as if he was from a different species, or something' (Miller, J. (1993). *The Passion of Michel Foucault* (New York: Simon & Schuster, p. 201–3).

8 See Smith, E. A. (2013). Agency and adaptation: New directions in evolutionary anthropology. *Annual Review of Anthropology* 42: 103–20, https://doi.org/10.1146/annurev-anthro-092412-155447. Things get rather confusing here. Behavioural ecology is an evolutionary approach. As Smith notes, evolutionary approaches have often been criticized by social scientists as allegedly not allowing for human agency, whereas 'sociocultural' approaches apparently do. I still maintain that behavioural ecology is agentic in the sense I am using it here, since it assumes individuals actively and plastically make efforts to pursue their survival and reproduction given the circumstances in which they find themselves. The individuals considered by behavioural ecology are thus not just agentic, but have something specific to be agentic about. In my view, whether a position does well or badly at allowing for human agency is orthogonal to whether it self-identifies as 'evolutionary'. Strongly culture-first positions struggle to recover an interesting or rich notion of agency, since they view individuals as basically dupes, uncritically susceptible to their

of Sociology and Anthropology are more cultural. Clearly this split is a problem. Some kind of agreed division of labour persisted for a few decades. For example, for large-scale societies, the discipline of Economics dealt with monetised, rational, instrumentally-motivated exchanges of goods and services between strangers, using agentic thinking, whilst Sociology dealt with non-monetised, normatively-governed features of life such as family formation, households, and voluntary associations, using, with exceptions, mainly cultural thinking. But this division of labour does not really work. For one thing, the rational-actor reasoning of Economics can be readily applied within households and to non-monetised interactions; for another thing, it is clear that morality and normative concerns play a big part even in monetised transactions amongst strangers.[9] When you are in a restaurant in a foreign city, do you pay a tip? If you are anything like me, your first question will be: what do people round here generally do? I want to do that.

Cultural and agentic thinking both capture something real, and satisfactory social theories need to encompass both. It's hard to find clear roadmaps for how to do this: usually authors start off in the camp belonging to one perspective, and at most hint at the other. The most useful source I have encountered is Herb Gintis' *Individuality and Entanglement*.[10] Gintis' book is a serious call to arms to reintegrate Economics with Sociology and Anthropology, and hence bridge the agentic and the cultural. Central to Gintis' analysis is the rational actor model. At the primary level, humans are agentic decision makers. Typical economist, you say, this is just the same old individualistic, agentic stuff. How do you explain morality? How do you explain why people follow norms when no-one is watching? Hang on—you are in danger of making a common conflation of *rational actor* with *self-regarding actor*. The micro-economists' rational actor model only says that people can be modelled as having preferences, and that when they do things, they follow these preferences in a consistent way. For example, if they prefer A over B and B over C, then they will choose A over C. The model

---

social and historical context. This is true whether such positions adopt evolutionary paraphernalia or not.

9   See Bowles, S. (2016). *The Moral Economy: Why Good Incentives Are No Substitute for Good Citizens* (Yale: Yale University Press).

10   Gintis, H. (2017). *Individuality and Entanglement: The Moral and Material Bases of Social Life* (Princeton: Princeton University Press).

nowhere says that the preferences people hold have to be selfish ones. The preferences could be aesthetically, morally or socially motivated. They could be altruistic or self-harming. The rational actor model only makes a claim about how preferences will be translated into decisions. Thus, it's a fallacy whenever you read (and you often do read): 'The facts that many people give blood, or donate money to strangers, violate the predictions of the rational actor model'. Not necessarily. The donors could be acting perfectly rationally, just as the rational actor model requires, if their preferences include improving the welfare of others. What generosity violates is the hypothesis that people's preferences are exclusively self-regarding. But you should have noticed that anyway.

Gintis suggests that we have three broad kinds of preferences, self-regarding (when we are hungry, we want food); other-regarding (when we see someone in distress, we want to help); and normative (we want to follow the rules, because they are the rules). But how do we integrate these very different preferences, when so often it seems that they might point us in different directions? There is no fundamental problem here. After all, we have no difficulty with the idea that we might integrate conflicting self-regarding preferences: we simply weigh them against one another, trading off if need be. I have to decide how much of my money to save, and how much to spend on running shoes. I might well be influenced in my trade-off by the prevailing rate of interest on savings, or the prospect of a looming expense next year, or a sale at the running shop. The same logic of weighing up and trading off applies when integrating self-regarding and non-self-regarding preferences. I want to help others, but the amount I invest in doing so will reasonably depend on the effectiveness of help I can deliver to them, the other calls on my time, and any rules I would have to break to do so. On my bicycle I like to stop at red lights, even when no-one is around—I think it's a good norm—but I might sacrifice obeying it if I was in a hurry, or in order to come to someone's rescue. And if I thought it was a bad norm, I would have a lower threshold for breaking it. So, in short, all of our varied motivations, self-regarding, other-regarding and normative, can simply be thought of as producing preferences of a unitary kind, preferences that get weighed against one another and traded off in moment-to-moment decision making, according to the context.

Because we have normative preferences—desires to do the appropriate thing, not for instrumental reasons, but for its own sake—a social and moral order can emerge, be largely obeyed and be somewhat stable, even when there are moment-to-moment self-regarding incentives for violating it at every turn. The order becomes internal and psychologically real. This sounds like a good thing, but is not always so. If the local norm seems to be that other people behave selfishly, then selfishness can spread rapidly and become a locally stable anti-social order.[11]

The existence of normative preferences allows Gintis to capture what is good about cultural thinking. However, our normative adherence is not slavish; because we also have self-regarding and other-regarding motivations, if the norms get too costly to our self-interest or the welfare of others, we may prefer to violate or abandon them. This captures what is good about agentic thinking. Thus, norms both *structure* our behaviour in the short term, and yet *are structured by* our behaviour in the long term, in the sense that bad norms can be changed.[12] Norms are empirically fixed, but transcendentally negotiable. And there is an emphasis I would like to add: when people find norms too costly, often they do not just individually abandon them. Instead or in addition, they talk to others about changing them. They produce, in the public sphere, reasons and arguments for why the social order should be different: political actions. Sometimes they even manage to persuade one another, and social change ensues. This aspect of social life, the use of reasoning and conversation as a means to change the rules of the games we live by, all the while accepting that there must be rules, is neglected in many contemporary treatments of culture, which focus instead on automatic, unreasoned socialization. But it is the possibility of reasoning and

---

11 Keizer, K., S. Lindenberg and L. Steg. (2008). The spreading of disorder. *Science* 322: 1681–85, https://doi.org/10.1126/science.1161405; Schroeder, K. B., G. V. Pepper and D. Nettle. (2014). Local norms of cheating and the cultural evolution of crime and punishment: A study of two urban neighborhoods. *PeerJ* 2: e450, https://doi.org/10.7717/peerj.450

12 For example: 'it would be wrong to think of [the social order] as a dictator who rules by force [...]. Social norms will not be followed when they are not considered legitimate [...]. Moreover, social norms generally are instantiated and changed through collective action, so that [the social order] itself is the product of a social will' (Gintis, H. (2017). *Individuality and Entanglement: The Moral and Material Bases of Social Life* (Princeton: Princeton University Press, p. 113).

conversation that gives me, as a scholar and a member of civil society, most hope for our common future.[13]

Gintis' analysis has the capacity to generate everything from extreme cultural thinking—if you assume that people's intrinsic preference for upholding norms is very strong relative to their other preferences—through to business-as-usual agentic analysis, if you assume that normative concerns have a trivially low weight compared to other concerns. The strength of normative motivation is thus like a slider with which you can fade from the very agentic to the very cultural and back again. Gintis does not solve where the slider is actually set for the typical human. Indeed, different passages of his book have quite different emphases in this regard. But surely, if we can take this as a framework, we can go beyond simply calling one or the other style of explanation *wrong*, and instead design a unified empirical programme that tries to find out where the slider is set, and what affects this setting.

§

I worry that we are too cultural when reasoning about people who are very different from us, and too agentic when reasoning about ourselves and our friends. Let's say anthropologist A claims, in her academic work, that the behaviour of the tribal people she studies is well explained by their propensity to internalize norms through socialization, leading to slavish within-group homogeneity in values and beliefs. Presumably, she doesn't believe these same principles account for her own behaviour in adopting this view. When asked, she doesn't answer that the reason she holds this view is because she was socialized to do so, and she slavishly accepts whatever she is taught. She says she adopted her view because she thought about it deeply and thinks it is a good theory, possibly in spite of the view being non-normative in the field. Her daily professional life shows theorist A that her fellow anthropologists don't easily accept the norms of the disciplinary community either: they take great pleasure in ceaselessly and idiosyncratically quibbling with her,

---

13 An approach very congenial to the perspective outlined in this paragraph has been developed by Simon Powers. He seeks to explain the form of social institutions by modelling strategic agents who interact economically in transactions following the current social rules, but also interact politically to change those rules for the future. See Powers, S. T. (2018). The institutional approach for modeling the evolution of human societies. *Artificial Life* 24: 10–28, https://doi.org/10.1162/artl_a_00251

for substantive as well as strategic reasons. As a consequence, far from there being cultural homogeneity in her discipline, there are as many shades of opinion as there are scholars, if not more. So the cultural group that theorist A actually has most experience of—her academic community—doesn't seem to fit the very theory she espouses. *She and her fellow academics* live in an agentic world, where different individuals have different values; beliefs require substantive justification; everyone is intensely sceptical about the claims of others; and the result is a myriad of clashing and shifting opinions within the same cultural group. *Her study subjects*, by contrast, seem to live in a cultural world where everyone credulously and automatically accepts what they are taught, and there is stable within-group consensus. So either: academics in Western societies are profoundly different from other humans; theorist A is right about her study subjects, but deluded about herself and her fellow academics; or theorist A is right about herself, but deluded about her study subjects.

It's easy to see where this double-think comes from. When we don't know much about a category of people, that category looks homogenous, and we represent it cognitively with a few sweeping and static generalizations. That's all our direct experience allows us to do. So that's how, if we are not very careful, we end up thinking about people in faraway places or distant times: more culturally. But where we have more direct personal experience, we build a richer representation, with more room for individual heterogeneity, the diversity of motivations, the within-group conflicts, and the ever-shifting dynamics. So that's how we think about ourselves and the other academics within our own discipline: more agentically. Thus, a bullshit test for social-theory frameworks that I rather like is to ask: do I find that framework rich enough to account for the social lives of the people I actually know? If the answer is no, then I see no reason it should be rich enough for anyone else either.

A version of double-think that I encounter in my own work concerns the behaviour of poor people, and why it differs from the behaviour of rich people. Poor people within developed countries are relatively likely to do various things that harm their health, such as smoking and overeating. A frequent mode of explanation for this in the literature is to say that, because their adverse lives grind them down, poor people are in less of a position to exert agency or free choice than rich people are, so they end up manipulated by the tobacco or sugar industries.[14] I find this argument uncomfortable. I accept of course that poor people have fewer options than the rich (there are fewer ways to furnish 3000 calories per day on a small budget than on a large one), and may also experience different benefits from health behaviour (no point in avoiding smoking if you are likely to be killed in an industrial accident by age 50 anyway). I also accept that people can be manipulated by commercial interests. But the argument that rich people are somehow more agentic than poor people, rather than equally agentic but with a different set of constraints and incentives, disturbs me. In effect, it seems to be saying, the rich may be fully human rational actors responsible for their decisions, but the poor are just credulous patients doing what they are told; a lower stage on some kind of *scala rationis humanae*. They need to be helped, as children must be, up to the point where they will be capable of choosing for themselves, but they are not yet there. The fact that these ideas come from scholars who are sympathetic to poor people, critical of capitalism, and progressive in intent, does not for me completely mitigate the discomfort of the double standard. I could live with the idea that we are all the passive victims of commercial interests; or the idea that we all choose our health behaviours, according to the constraints and incentives of our circumstances. But I can't get entirely comfortable with the idea that there are different modes of explanation for rich and poor.[15]

---

14 For example: 'poor and food-insecure groups have the least agency to resist commercial interests [...] this lack of agency is itself promoted by corporate manipulation of dietary quality and food availability': Wells, J. C. K. (2017). Obesity is not just elevated adiposity, it is also a state of metabolic perturbation. *Behavioural and Brain Sciences* 40: e105, p. 35–6, https://doi.org/10.1017/s0140525x16001552. See also Marmot, M. (2015). *The Health Gap: The Challenge of an Unequal World* (London: Bloomsbury) for similar views.

15 In a recent discussion, Gillian Pepper and I were accused of subscribing to the 'poor but neo-classical' style of analysis. This style of analysis basically says that poor people follow the same neo-classical micro-economic principles as anyone else. On balance we take it as a compliment: we're not sure we are really neo-classical, but if we were to be, we would want to be neo-classical about rich and poor alike. See Carmel and Leiser's commentary (p. 22), and Pepper and Nettle's response (p. 45–54), in Pepper, G. V. and D. Nettle. (2017). The behavioural constellation of deprivation: Causes and consequences. *Behavioral and Brain Sciences* 40: e314, https://doi.org/10.1017/s0140525x1600234x

What I take away from this is the following principle: we should have a strong prior that *all humans are just as cultural and just as agentic as each other*. I don't mind where you place your thinking on the cultural-to-agentic continuum—though I am more disposed to the agentic than the cultural, I can see the merits of several different positions. Wherever you place it, though, I think it should start at the same point for everyone, regardless of: the colour of their skin; their level of education; whether they are a hunter-gatherer, a subsistence farmer, an unemployed welder from Glasgow, a university professor, an antique Roman, or a Dane. Either we are all agents, it seems to me, or none of us is.

4. What is cultural evolution like?

Different problems require different tools

–Tim Lewens[1]

For over twenty years, I have been having an on again/off again affair. The other party in the affair is appealing and effortlessly elegant. She promises to wash away the mundane, equivocal, hesitant, mutually contradictory strata of my work as a behavioural scientist and replace them with something simpler, crystalline and more powerful. I encounter her from time to time in the course of my professional duties. We write a paper or two together. I expect summits at Camp David. But within a year or two I start to pull away. She is brittle behind the mask. I start to worry about how the bills will really get paid. When the mirrors have stopped dazzling and the smoke has cleared, I realise I still have all the problems I had before.

The other party in the relationship is an idea. Actually, two linked ideas: (i) that cultural change is a Darwinian process; and (ii) that because (i) is true, social science can be substantially simplified under the rubric of a single body of theory that does the same job, in the same way, as evolutionary theory does for genetic evolution. These ideas have been knocking around for about forty years. They have their passionate adherents.[2] But they continue to attract scepticism, and despite all the conceptual discussion, I don't notice journals of sociology, politics, social anthropology, history, cultural studies and so on being full of empirical calculations of cultural fitness, cultural relatedness, cultural heritability

---

1   Lewens, T. (2017). *Cultural Evolution: Conceptual Challenges* (Oxford: Oxford University Press, p. 146).

2   Notable adherents are: Dawkins, R. (1976). *The Selfish Gene* (Oxford: Oxford University Press); Mesoudi, A., A. Whiten and K. N. Laland. (2006). Toward a unified science of cultural evolution. *Behavioral and Brain Sciences* 29: 329–83, https://doi.org/10.1017/s0140525x06009083; Mesoudi, A. (2015). Cultural evolution: A review of theory, findings and controversies. *Evolutionary Biology* 43: 481–97,

    https://doi.org/10.11647/OBP.0155.04

and so forth, in the way that journals of evolutionary biology are full of the genetic versions of these notions. So what is going on?

There are a number of possibilities. One is that ideas (i) and (ii) are fundamentally correct, and Rome wasn't built in a day. The history of science shows us that the right idea takes a long time to rise up through layers of inertia, tradition and disciplinary resistance. Plate tectonics, for example, took about 60 years from first, derided claims to universal acceptance. The acceptance took the form of a characteristic S-curve: very slowly rising for a long time, then a phase of rapid spread, then the slow mopping up of the few remaining non-believers. So as a cultural Darwinian, you must tell yourself that you are just entering the accelerating phase on the S-curve; *this year*. My problem is: I thought that twenty years ago, when I wrote my first cultural Darwinian papers.[3] I am still waiting.

Another possibility is that idea (i) is wrong, and hence idea (ii) also fails, but there are other reasons people cling to them. Research in the humanities and social sciences is in slow decline (not without a fight). Eighty years ago, to a fairly reasonable approximation, the humanities and social sciences were what universities did. Today, also to an approximation, universities have a dual role: they teach students in humanities and social sciences, and they do research in biology. A glance at the difference in teaching load, and research and infrastructure funding, between my faculty in my university (Biomedicine), and the Faculty of Humanities and Social Sciences, is instructive. We are researchers with expensive labs and technical support teams who give the odd lecture; they are teachers who occasionally manage to scrape

---

https://doi.org/10.1007/s11692-015-9320-0. Before I annoy anyone any more than I need to, I should distinguish between cultural evolutionary thinking, and cultural Darwinism. The former is simply the attempt to understand the population-level consequences over time, for human societies and their cultural attributes, of individual patterns of learning and cognition. It is broader than cultural Darwinism, a subset of cultural evolutionary thinking which sees cultural change as a process of Darwinian selection and hopes through that insight to radically transform the social sciences. Cultural evolutionists are not necessarily committed to cultural Darwinism: see Lewens, T. (2017). *Cultural Evolution: Conceptual Challenges* (Oxford: Oxford University Press) for discussion.

3   For example Nettle, D. (1999). Functionalism and its difficulties in linguistics and biology. *In Functionalism and Formalism in Linguistics* (M. Darnell et al. eds, Amsterdam: Benjamins, p. 445–62), https://doi.org/10.1075/slcs.41.21net

the time to write articles and books. Anything that seems to offer the humanities and social sciences the possibility of getting what biology has had seems worth grasping at. So idea (i) does not quite die. Sadly, those who want to save the humanities and social sciences through a Darwinian theory of culture are probably looking in the wrong place. The recent growth of biology is almost entirely in cellular and molecular work, a part of biology largely free from the guiding light of Darwin's dangerous idea. The driving forces have been rapid technical progress in what researchers can measure, and the computational firepower to mine the resulting big data. So that's where a lot of the smart money in social science is going to go too.

As so often in life, I find myself somewhere in the middle ground. The analogy between genetic and cultural evolution is strong enough that it continues to capture my theoretical attention.[4] On the other hand, it's not straightforward enough for ideas (i) and (ii) to get off the ground in a major way. I don't expect the revolution imminently. Hence my on/off affair. Hanging on to both edges, as usual.

§

Genetic and cultural evolution are not exactly isomorphic. Everyone admits that. On the other hand, there are some general similarities: something gets transmitted from individual to individual; some things spread and others become extinct; there is a kind of descent with modification, and so on. So the issue is: what do we do with this partial similarity? We could either: define Darwinian processes rather narrowly, and thereby include genes but exclude culture; or find broader ways of defining Darwinian processes, so as to include the cultural case as well as the genetic.[5] Clearly the answer we get to the question of whether cultural change is Darwinian will depend on the definition of 'Darwinian' we adopt. A more fruitful avenue, to my mind, is to ask: what special job does evolutionary theory do for organismal biology,

---

4   Most recently in El Mouden, C. et al. (2014). Cultural transmission and the evolution of human behaviour: A general approach based on the Price equation. *Journal of Evolutionary Biology* 27: 231–41, https://doi.org/10.1111/jeb.12296

5   As in Claidière, N., T. C. Scott-Phillips and D. Sperber. (2014). How Darwinian is cultural evolution?. Philosophical Transactions of the Royal Society B: *Biological Sciences* 369: 20130368, https://doi.org/10.1098/rstb.2013.0368

and what are the properties of genetic evolution in virtue of which it can do that job? Then we can assess the extent to which cultural evolution has those properties, and hence whether a Darwinian 'cultural evolutionary theory' could do that job. I'm in an off-again phase, and so I am going to conclude that cultural evolution generally lacks the properties and hence 'cultural evolutionary theory' (thought of in this particular way) can't really do the job. This much has often been said before, by better people than me. Perhaps I have a slightly more unusual insight, though, which is that the real problem for the hope of a unified cultural Darwinian theory is that different cultural cases are very different from one another, and hence approximate the genetic situation to different degrees. This is a serious blow to hope (ii), the hope of simplification of the social sciences under a cultural Darwinian banner.

It's a commonplace that you have Darwinian evolution whenever you have variation (different individuals in a population have different traits); heredity (offspring resemble their parents); and differential reproductive success (the descendant generation differentially samples from the ancestral one, or equivalently, different individuals have different chances of becoming ancestors). This much is true, but I think we need to build the requirements up more slowly. First, there must be a clearly defined population of *individuals* through time; you need to know what your individuals are. Second, within that population, you must be able to identify which individuals are *descendants* of which others, and which are not. Without being able to do this, there is no hope of measuring reproductive success, since the very notion depends on descendant-counting. Third, these individuals need to have *traits*: characters, discrete or continuous, that you can measure, and hence characterize straightforwardly the extent to which descendant is like ancestor.

With these requirements in place, we can characterize the way any particular trait changes from one generation to the next. This was famously done by George Price, in the Price equation.[6] So general and important is this equation that it has a movie based on it, $w\Delta z$ (directed by Tom Shankland, 2007; apparently it's a horror story). The Price equation says, in words, that in each generation:

---

6   Price, G. R. (1970). Selection and covariance. *Nature* 227: 520–21, https://doi.org/10.1038/227520a0

Total evolutionary change in the trait =

A bit due to selection +

A bit due to average transformation

The Price equation also tells us how to compute the value of each bit. The bit due to selection is exactly the covariance between the value of the trait, and *fitness*, where fitness is the individual's number of descendants in the next generation, divided by the population average number of descendants. A covariance is like a correlation: it can be positive, negative or zero. So let's say that the trait is nose length. If it is the case that the longer your nose, the higher your reproductive success on average, then the covariance between nose length and fitness is positive, and the value of the bit due to selection is positive. This means selection is making noses longer from generation to generation. If longer noses tend to be associated with reduced reproductive success, then the covariance of nose length with fitness is negative, and hence selection is making noses shorter. And of course, the length of your nose may have no systematic relationship with reproductive success, in which case, the bit due to selection has a value of zero, and there is no directional selection on the trait.

Then there is the bit due to average transformation. Imagine a case where, because of some strange quirk of genetics or development, offspring always had noses that were a bit longer than the average of the lengths of their two parents' noses. It's easy to see that noses would get longer over evolutionary time, even in the absence of any natural selection. In fact, they could get longer over evolutionary time even with some natural selection acting in the opposite direction. The Price equation tells us exactly when this will happen: when the average amount by which an offspring's nose length exceeds those of its parents (the bit due to average transformation) exceeds the negative covariance between nose length and fitness (the bit due to selection). This is because, to get the total evolutionary change from the Price equation, you simply add the two bits on the right-hand side together.

It's important not to confuse random mutation or imperfect heredity with average transformation. Let's say there is quite a lot of genetic mutation, so that offspring nose length is not perfectly predicted by parent nose length. If an offspring's nose is just as likely to be a bit shorter than their parents' as a bit longer than their parents', then *on average*, their nose is neither longer nor shorter, and so the bit due to average transformation is still zero. The genetic mutation averages itself out as it were, and the total evolutionary change comes from the bit due to selection alone, even though individuals do not perfectly resemble their parents.

§

With this exposition in the bank, we can begin to ask what properties genetic evolution has that allow it to do a special job for organismal biology. And it does do a rather special job. A relatively straightforward mode of theorising, in which trait evolution is explained on the basis of higher or lower fitness, can be brought to bear in much the same way on any organismal trait, be it the dimensions of the hummingbird, the shape of fish eyes, the propensity to help others reproduce, or the way animals forage. The mode of theorising can be used regardless of what the trait in question actually is, and in particular, without knowing anything about the details of the molecular genetic mechanisms involved. Why can we do this?

The first reason we are able to do this in the genetic case is that the relations of ancestry and descent are straightforward. I have just two genetic parents. No one else has influenced the length of my nose (heritably, that is). Those same two individuals are my parents in respect of all of my other traits, not just nose length. The ancestor-descendant link points in one direction only: I can't back-influence the heritable traits of my genetic parents. And how many parents I have does not depend on the lengths of their noses. That sounds bizarre, but is not guaranteed in the cultural case. For example, I might sample the way of life of the first few people I encounter. If it seems to suit me, fine, I follow it, but if it seems dreary, I might go looking for other people to emulate. This is me shopping for cultural ancestors on the basis of the traits they offer, something we don't get to do with our genes. Because of the straightforwardness of the ancestor-descendant mapping in the genetic case, you can readily count offspring and measure reproductive success. And then it's easy to compute the value of the bit of evolutionary change due to selection: measure the trait you are interested in, count descendants, apply the formula for a covariance.

The second reason fitness is central to organismal evolution is that for most traits considered by biologists, the bit due to average transformation is zero or negligible. The conclusive evidence for this is that when, in $w\Delta z$, the evil serial killer carves the Price equation onto the flesh of her victims, mindful of parsimony and apparently having the biological case in mind, *she only includes the term for the bit due to selection on the right-hand side.* She can simply leave out the term for the bit due to average transformation because, usually, genetic reproduction systems have no particular transformational drive one way or the other. Thus the Price equation, for genetic cases, usually reduces to evolutionary change being equal to the covariation between the trait value and fitness.

Why is average transformation negligible as a source of change in genetic systems? It is because, in some profound sense, the function of DNA replication mechanisms is to indifferently reproduce whatever is thrown at them. That's their job. DNA replicase is indifferent whether it replicates a cytosine or replicates a guanine, indifferent indeed to what if anything the particular stretch of DNA it is currently copying actually does. It simply *has no interests* other than to fulfil its evolved role of making DNA into more DNA. To the extent that mutation happens (and it does, though overall fidelity is high), this mutation can be fairly analogised, as it often is, to 'mistakes' or 'imperfections' in the replicative process. And at reproduction, fair meiosis generally ensures that no variant gets a leg up, on average, over any other.

The profound content-indifference and impartiality of DNA replication provides us, as scientists, with the option of abstracting away from a lot of the details of how replication and reproduction actually work in each particular case. You can usefully treat genes in populations as if they were simply beans being drawn from a bag.[7] To think about the evolution of nose length, you don't really need to know about the molecular details of which stretches of DNA influence nose length and how, at least in the first instance. This is because you can take it for granted that those molecular details, however they work out, come down to offspring of long-nosed parents having long noses, plus some effectively random noise. So we can make a great deal of progress just by knowing that nose length is heritable, and measuring

7   Haldane, J. B. S. (1964). A defense of beanbag genetics. *Perspectives in Biology and Medicine* 7: 343–59, https://doi.org/10.1353/pbm.1964.0042

how it correlates with fitness, without getting bogged down in the messy biology of the specific case. Indeed, the messy biology of the mechanisms in each particular case is still largely unknown to us: a black box that we have scarcely as yet peered inside. And because in each biological case, we have been able to abstract from the details of the molecular and developmental processes involved, then all normal biological cases are effectively like one another from an evolutionary theoretical point of view. Transmission is by fair replication from one or two parents, and selection (plus drift) are, to a first approximation, the drivers; the rest is about the impact of the trait on fitness in populations. The general formulae apply. This is the sense in which we can sensibly talk of *an* evolutionary theory, rather than one theory for the evolution of eye shape, based on the developmental biology of eyes, one theory for the evolution of blood proteins, based on the physiology of blood proteins, and so forth.

§

Now let's start to think about cultural cases. We start where I started my career, with the evolution of words.[8] Consider Zipf's law, which states that words that are used more frequently in a language tend to be shorter, whilst rare words are longer.[9] This looks pretty much like an adaptation: it benefits speakers in terms of overall articulatory effort if the shortest available word forms are used up on the meanings we need most often, and longer words forms saved for meanings we don't need to utter very often. And it's tempting to characterise the process producing the pattern in terms of selection. Doing so, in fact, goes right back to Darwin, who noted in *The Descent of Man*:

> As Max Muller [...] has well remarked: — "A struggle for life is constantly going on amongst the words and grammatical forms in each language. The better, the shorter, the easier forms are constantly gaining the upper hand, and they owe their success to their own inherent virtue."[10]

---

8   Nettle, D. (1995). Segmental inventory size, word length, and communicative efficiency. *Linguistics* 33: 359–67, https://doi.org/10.1515/ling.1995.33.2.359

9   Zipf, G. K. (1949). *Human Behavior and the Principle of Least Effort* (Cambridge, MA: Addison-Wesley).

10 Darwin, C. (1871). *The Descent of Man, and Selection in Relation to Sex* (London: John Murray, p. 465–66).

Let's unpack how this might work. People have a set of cultural parents from whom they learn their word forms. We can assume that they learn their word forms early in life, and the forms are fixed thereafter. So that means word-form learning really is a bit like genetic inheritance, with ancestor-descendant arrows pointing in one direction only, from older people to younger. The set of cultural parents is broader than just their genetic parents of course, but the Price equation can be generalised to an arbitrary number of parents, and indeed can be generalised to a case where parenthood is a matter of degree: you have many cultural parents, and some are more influential than others.[11] Fitness then becomes not the number of descendants you have, but the average strength of your influence on all the individuals in the next generation. No intrinsic problem there; though, given that the set of people you learn from can be different for different traits (I learn playground games from my peers, science from my teachers), it does follow that every individual has not just one cultural fitness, but indefinitely many cultural fitnesses, one for each cultural trait.

The main quibble with analogising the emergence of Zipf's law to adaptation through natural selection is that it is not clear the adaptation arises from the selection bit of the Price equation, rather than the average transformation bit. It could be that as you grow up, you hear various idiosyncratic variant word forms spoken around you, and you have a bias towards adopting in your own speech those variants that give you short word forms for frequent meanings. That would be a kind of selection. But an alternative (not mutually exclusive) mechanism is that as you use language, you tend to spontaneously shorten words or phrases you utter frequently. You might do this even if you have not heard the people you learn from do so. Geddit? This could be the source of Zipf's law. Word forms tend to start life long, and if they are used often, speakers spontaneously contract them through economy of effort. If that's right, then the adaptive change is not actually due to *selection*, but rather a particular bias in average *transformation* (a person's habitual word form for a common meaning will be a bit shorter than

---

11   See El Mouden, C. et al. (2014). Cultural transmission and the evolution of human behaviour: A general approach based on the Price equation. *Journal of Evolutionary Biology* 27: 231–41, https://doi.org/10.1111/jeb.12296

the average length of that word form in the models from whom that speaker learned).

Whether word evolution is due to transformation or selection is, you might say, a rather unimportant technicality. Perhaps. But there have been very fine experiments in recent years using so-called *transmission chain* or *iterated learning* paradigms. Here, one participant learns something (a starting stimulus furnished by the experimenters); a second participant learns from the first; a third participant learns from the second; and so forth. These experiments have been applied to cultural content as varied as stories, communication conventions, rhythmic patterns, statistical relationships, and many other things.[12] They show in fascinating detail how cultural change can be rather fast, and decidedly non-random: typically, across just half a dozen links of the chain, the learned content changes in patterned ways. The contents go from being essentially random in the first generation to having the rich structure seen in real cultural representations: communication systems become grammatically regular; rhythms acquire the regular pattern of strong and weak beats you see in all music; and complex random statistical scatters are reduced to simple, memorable stereotyped relationships.

I love these experiments. They show us how culture evolves. But there is no selection going on. There can be none by design, since every participant has exactly one cultural ancestor and exactly one cultural descendant. The bit due to selection in the Price equation is therefore exactly zero. The rich, non-random shaping of the content that we can see in these experiments must be entirely due to average transformation — the way that human participants actively shape the information they are exposed to, in accordance with their purposes, strategies and biases — and not at all due to selection. That's a key difference from genetic evolution. And it raises the question: if you can get good experimental analogues of cultural change in an experimental set-up that excludes

---

12   Kirby S., H. Cornish and K. Smith. (2008). Cumulative cultural evolution in the laboratory: An experimental approach to the origins of structure in human language. *Proceedings of the National Academy of Sciences of the USA* 105: 10681–86, https://doi.org/10.1073/pnas.0707835105; Ravignani, A., T. Delgado and S. Kirby. (2017). Musical evolution in the lab exhibits rhythmic universals. *Nature Human Behaviour* 1: 0007, https://doi.org/10.1038/s41562-016-0007; Griffiths, T. L., M. L. Kalish and S. Lewandowsky. (2008). Theoretical and empirical evidence for the impact of inductive biases on cultural evolution. *Philosophical Transactions of the Royal Society B: Biological Sciences* 363: 3503–14, https://doi.org/10.1098/rstb.2008.0146

any selection, to what extent is selection needed to do the explanatory work for cultural evolution in the wild?

§

They say hard cases make bad law. But there ought to be a similar adage for easy cases and bad generalizations. By choosing the cultural cases that best fit the analogy between genetic evolution and cultural change, we overestimate how good the analogy is overall. In the example of word forms considered until now, there are some rather atypical circumstances that obtain. It is reasonable to assume that one learns one's word forms early in life, and hangs on to them thereafter. So the ancestor-descendant links go in one direction only, from older to younger people. But this is not the general case. Consider the cultural evolution of scientific ideas. I have learned a lot about science from my long-term collaborator Melissa Bateson. But I have also transformed what I have learned from her, using my own particular cognitive operations, and back-influenced her in turn. And then of course she has reflected on and transformed those ideas still more, influencing me again. So who is ancestor and who descendant? How will we deal with this if we wish to maintain some parallel between genetic evolution and cultural change?

We could say that a person becomes a new individual in respect of any particular cultural trait every time they have a change of idea. So Melissa Bateson of 2017 is a cultural descendant of, among others, Daniel Nettle of 2016, in the domain of scientific ideas. But Daniel Nettle of 2016 is a cultural descendant of Melissa Bateson of 2015. So we would have to admit that Melissa is one of her own cultural grandparents in the domain of scientific ideas. Maybe that's ok. But it means that, for culture, Melissa does not just have indefinitely many different fitnesses. She is also indefinitely many individuals (some of whom are ancestors to some of the others in some domains). And if that weren't complex enough, many of those individuals are alive at the same time. She wrote articles twenty years ago that are still influencing biological ideas in January 2017 via a route other than what she believes in December 2016 (she may even have forgotten what they say). I hope it's clear that the moment you have the possibility of repeated and continuous learning over the life-course (let alone literacy), the whole question of ancestry, descent, and fitness

becomes really rather difficult to track; and culture has potential dynamics
that cannot be captured with models inspired by beanbag genetics.[13]

§

Let's take another example close to my heart: the cultural evolution of
my local cross-country running league. If cultural Darwinism can do
anything, I really want it help me here, because this venerable institution
shows clear evidence of descent with modification, plus evidence of
adaptation to human purposes, over the course of its 120-year history.[14]
Like other Victorian running leagues, it began with 'hare and hounds'
pursuit races. (This is the origin of the name 'harriers' in the names of
many running clubs in the English-speaking world, and also the source
of the term 'steeplechase': routes were not fixed, but a small number of
'hares' would head across the countryside for visible landmarks like
church steeples, jumping streams and fences, pursued by the pack of
'hounds'.) The runners in the pursuing pack were not racing against
each other for most of the run. They would deliberately stay together
in a peloton until a final competitive sprint for around the last mile,
whose sudden onset was orchestrated by a special runner called the
Whipper-In. After 1950, this system was simplified to 'all out' racing:
the participants all competed against one another from gun to tape, not
just for the final sprint; the hares and Whipper-In were abandoned.

A subsequent innovation was a handicap system, whereby runners
who finish high up in one race must start with a handicap in subsequent
races of the season. This keeps things challenging for the fastest
people, whilst providing some hope for the slower ones. The handicap
system applies only to league matches—there is also one cup match
a year, where everyone goes off together. The rules for promotion to
a higher handicap in league matches are not symmetrical with those
for demotion: you can increase your handicap after every match, but
only decrease it at the end of the season. This means that the first group
to set off gets smaller and smaller as the season goes on, as more and
more people attract handicaps. Thus, people who would have no hope

---

13   See Strimling, P., M. Enquist and K. Eriksson. (2009). Repeated learning makes
     cultural evolution unique. *Proceedings of the National Academy of Sciences of the USA*
     106: 13870–74, https://doi.org/10.1073/pnas.0903180106

14   The authoritative source is Jenkins, A. (2016). *Whipper-In: The Northumberland and
     Durham Paperchase League, the Early Years, the Forerunner of the North East Harrier
     League* (Alnwick: Wanney Books).

of winning in the first match of the season do so in the last. Then there are complex rules for how the team's score arises from the positions of individual runners (it's a team sport); and, a very recent innovation, there are several divisions for teams, with rules for team promotion and demotion from their division.

I could go on, but I hope you can see that there is an exquisite cultural order here that has emerged by the slow reshaping of institutional tradition to conform to human needs and desires. Is there any way we can fit this slow reshaping into a Darwinian framework? It's not clear that there is variation and selection, since there has only ever been one cross-country league in Northeast England. So there aren't really competing variants with higher or lower fitness. After all the whole point of an institution is that at any given time, all individuals have to sign up to exactly the same rules. I suppose we could say that in some sense the rules of cross-country are in competition with other things people could spend their Saturday afternoons doing; perhaps the rules of cross-country have a fitness relative to golf, say, or gardening. But this is rather different from the genetic evolution of nose length, where noses of a given length are in competition with simultaneously-present noses that are slightly shorter or longer, driving nose length up an adaptive gradient of nose length.

And then there is how change works. There's modification, but is it usefully thought of as inheritance plus mutation? There is a league committee, who are bound by a written constitution. The committee sits down at the end of the season, looks at what went well and what people complained about, and uses reason and argumentation to decide whether there is anything they want to change. So you could perhaps say that there is a kind of virtual variation and selection process, whereby the committee simulate in their minds various alternative possibilities, then choose the one that seems best for the sport. That would be rather evolution-like in way, with the exceptions that the variants with lower fitness never get to actually exist outside of committee meetings.

Given that there is only one set of rules, the ancestor for each rule in each season seems to be the corresponding rule in the previous season—faithful inheritance, with agreed rule changes playing the role of mutation. But maybe that's not right. For example, the recently-introduced divisional system for teams, with its rules for promotion and demotion, is clearly similar to the rules that do the same job for the Football League. So perhaps the cultural trait we should be thinking about is not 'the rule system for cross-country running' but 'divisional systems in competitive sports'. Then we could talk of the fitness of divisional systems being high, as they have colonized new sports like cross-country running. But there might be things in some sports which are *a bit like* divisional systems, fashioned somewhat in the style of such systems, but not exactly the same as them. How would these contribute to the cultural fitness of the people who advocated a divisional system for the Football League? Would they count towards it, or not?

There are harder cases still. I fully expect within a few years, there will be pressure for women and men to run the same course (currently, senior women run two laps and senior men three). If this happens it will reflect the broader social concern about gender equity that we currently see, for example, in the debate about wages in the UK. Harriers bring that broader cognitive and political framing into their leisure activities. If gender-equalization happens, then we are effectively saying that the value of a trait in cross-country running (course length) has been influenced by the existing trait-value, plus a general concern about gender equity that is otherwise manifest in entirely different domains of human activity. This has no parallel in genetic evolution. It is as if you said: the two heritable influences on the length of my nose are the length of my father's nose, and my mother's sense of humour. But culture is pervasively like this. Cognition ranges promiscuously across domains and activities, seeking partial resemblances and relevant reasons, recombining, tidying up, reconstructing one thing in the light of another. This makes even identifying what the traits under cultural evolution are, as well as delineating who is culturally ancestral to whom, very complex at best.[15]

§

It is time to limp, bloodied but still determined, towards some more general conclusions. For genetic systems, transmission is achieved by

---

15 Claidière and colleagues provide a way of addressing some of this through their notion of hetero-impact: a feature in one cultural generation can have a causal impact on a different feature in the next. See Claidière, N., T. C. Scott-Phillips and D. Sperber. (2014). How Darwinian is cultural evolution?. *Proceedings of the Royal Society B: Biological Sciences* 369: 20130368, https://doi.org/10.1098/rstb.2013.0368

content-indifferent replication mechanisms; and the relationships of ancestor and descendant are straightforward in all cases, and the same for all traits. This allows us to bypass the details of how transmission is actually achieved, and go straight to a special type of explanation in which fitness, and the relationship of different organismal designs to fitness, do all the explanatory work. Evolutionary biologists have special names for the thinking that underlies this special type of explanation: beanbag genetics (idealizing genes to understand their dynamics in populations); ultimate reasoning (thinking about fitness consequences of a structure rather than the mechanisms that make it); the phenotypic gambit (ignoring the genetic architecture of the traits under study); and the behavioural gambit (assuming brains can deliver whatever is good for fitness without worrying how they might do so). And because you can do these special types of thinking for all traits, there is a coherent sense in which there is *one* evolutionary theory: one tightly-integrated, portable system of tools for working out change in any genetic system. The setting aside of the details of how transmission actually works is only a provisional strategy; and the worry is often voiced within biology that you can't entirely get away with it. In the end, the details of the available mechanisms are probably going to matter for what happens.[16] Still, people have been able to do a lot by making the idealizations and thinking at the ultimate level.

In the cultural case, transmission is achieved by human action and human thought. Humans are very far from indifferent about the contents of their acts and thoughts. The function of DNA replicase is to replicate DNA, but the function of humans is not to replicate culture. The function of humans is to be humans. This is a crucial difference. It means humans have all kinds of characteristic interests, strategies, goals, biases, priors, intuitions, and so forth. They apply shaping forces to whatever they transmit, sometimes unconscious and automatic ones, sometimes deliberate and reasoned ones; sometimes through individual action, sometimes through institutions. If it seems like humans unreflectively replicate just whatever society is doing, that's only because we focus on atypical cases. In the case of which side of

---

16 See for example Fawcett, T. W., S. Hamblin and L. A. Giraldeau. (2012). Exposing the behavioral gambit: the evolution of learning and decision rules. *Behavioral Ecology* 24: 2–11, https://doi.org/10.1093/beheco/ars085

the road to drive on, for example, my *only* interest in choosing one is to choose the same one as everyone else in the country. So for that trait, humans are remarkably obedient replicators of the society they live in, and a basically arbitrary norm is stable indefinitely. But most traits are not like that. Mostly, I have interests and biases that go beyond the mere desire for my behaviour to be the same as everyone else's.[17]

The shaping forces in culture will be different for every trait. So too will the relationships of ancestor to descendant: you learn different things in different ways. So too will the dynamics: you learn some things once and for all; others you continuously update through your life. There is thus no very general expectation we can form, for example that humans will hold such beliefs as maximise their cultural fitness, or anything like that. This is not a counsel of despair. I am not just saying 'it's all very complex', or that we are limited to a kind of post-hoc or qualitative historical interpretation of cultural change, without hope of bringing it under the umbrella of natural science. Some of the transmission experiments described above show that this is not so. We can still formulate and test explanatory causal principles for the properties and dynamics of human cultures. But we need to begin from an appropriate framing of the problem in order to do this. I don't think that the failure for the 'beanbag' move to become widespread for culture in the way it did for population genetics is due to the intransigence or perversity of social scientists. I think it is due to substantive and interesting differences between the genetic and cultural cases.

So am I saying, in short, that there can be no such thing as cultural evolutionary theory? There can certainly be cultural evolutionary *theories*. One can—indeed, one must—model cultural regularities and cultural change as the population-level emergent consequences arising from the ways individual people learn, communicate, influence one another, think, remember and forget. And formal, computational and empirical tools are required in order to do this. Much of the scholarly enterprise known as 'cultural evolution' is simply the attempt to provide these tools, without necessary commitment to the idea that cultural change is a narrowly Darwinian process. My worries here leave this work and its motivation intact. But note the difference from the genetic evolution case.

---

17   For an approach to cultural evolution in whose spirit this paragraph was written, see Morin, O. (2015). *How Traditions Live and Die* (Oxford: Oxford University Press).

Whereas in genetic evolution, empirical knowledge of the transmission mechanisms in each case can be bypassed, in the cultural case, empirical knowledge of the transmission mechanisms in each case is precisely what you need to have in order to be able to build your explanation. You can't possibly understand how the repeated application of human cognitive biases shapes supernatural ideas without first studying in detail what those cognitive biases are. And your theory will only be as good as your characterisation of the cognitive or interactive processes that are doing the explanatory work. Theory is therefore dependent on knowledge of the transmission mechanisms in each individual cultural case, in a way that it is (arguably) not for genes.

It follows that there will be as many cultural evolutionary explanations as there are domains across which human cognition and human interaction are different. Thus, there can be *a* cultural evolutionary theory only in a weak sense, meaning the general set of recipes used in such explanations. These recipes will actually be rather varied, and will only work when made up with fresh ingredients from empirical psychology, cognitive science, politics, sociology, and so forth. This is rather different from the somewhat stronger sense in which there is *an* evolutionary theory in organismal biology. And that means if we wish to unify the social science disciplines, which surely we must wish to do, we must develop a slightly different banner under which to do it. I would put human action, or more generally human cognition, rather than cultural selection, at the heart of that enterprise.

I worry that I have spent a lot of this meditation being rather negative, so perhaps I should end instead by being positive: about humans, and about culture. Culture is not like DNA. It is the residue of past cognition and past interaction; a residue that is available to cognitively complex, socially interacting, purposeful, reasoning beings. And we are really good at doing things with this residue. The operations we perform on it are not limited to its reproduction. We can perpetuate it, yes, but also extend it, adapt it, discuss it, contest it, refuse it, restructure it, or redesign it; not in its interests, but in our own. The human capacity for purposive agency using the raw material embodied in culture is the resource a sensible progressive politics needs to be built upon. And it is the capacity that makes human culture, and indeed human beings, extraordinary.

5. Is it explanation yet?

> What we say sounds like an explanation—but really it is a
> terrible jumble that we are making up as we go along.
>
> –Nick Chater[1]

One of the most devastating rejoinders you can give an academic is to
characterise what they have offered you by way of an explanation as no
more than a re-description of the phenomenon at hand. For example,
let's say I am interested in the knotty problem of why people of lower
socio-economic position are less likely to successfully quit smoking
compared to those of higher socio-economic position.[2] I could offer you
the insights of the theory of planned behaviour, one of the most popular
theories in this kind of area.[3] The theory says (very roughly) that people
do healthy things when they want to do them, they think they should
do them, and they think they can do them. So perhaps the reason
people of lower socio-economic position are less likely to successfully
quit smoking is either (a) they don't so much want to; (b) they don't so
much think they should; and/or (c) they don't so much think they can.
All of these are testable: I could go off and ask a load of people, and
come back with some results. Let's say, hypothetically, I find that it's
mainly (b). Hurrah, I say, I have now explained the social gradient in
smoking cessation—in terms of a social gradient in the belief that it is
normatively desirable to cease smoking.

Here's where your wounding rejoinder comes in. All you have done,
you say, is to re-describe the social gradient of interest—poorer people
are less likely to quit smoking—as a social gradient in the extent to

---

1   Chater, N. (2018). *The Mind is Flat: The Illusion of Mental Depth and the Improvised
    Mind* (London: Penguin, p. 28).

2   Kotz, D. and R. West. (2009). Explaining the social gradient in smoking cessation:
    It's not in the trying, but in the succeeding. *Tobacco Control* 18: 43–6, https://doi.
    org/10.1136/tc.2008.025981

3   Ajzen, I. (1991). The theory of planned behavior. *Organizational Behavior and Human
    Decision Processes* 50: 179–211, https://doi.org/10.1016/0749-5978(91)90020-t

      https://doi.org/10.11647/OBP.0155.05

which people believe they should quit smoking. But where does *that* social gradient come from? Perhaps people of lower socio-economic position don't feel so much normative pressure to quit smoking because fewer of them successfully do so…but wait, isn't that where we started? As so often in social science, we have end up at a place where the thing being invoked to do the explaining (the *explanans*), does not seem entirely independent of the thing we wish to explain (the *explanandum*). And, more pressingly, you want to ask: where the hell did the *explanans* come from anyway? What explainans that? (Sorry.) To account for one pattern, I offered you another, but that other one seems immediately to cry out for a deeper explanation, an explanation that stands entirely free of the phenomena we are studying.

This is about the point where people like me, who advocate evolutionary, *aka* behavioural-ecological, explanations for patterns of human behaviour, pipe up. What we tend to say at this point is something along the lines of: what you other social scientists offer is some kind of *proximate* explanation for the phenomenon at hand: another phenomenon that stands immediately prior to the original one in the chain of causation. That's fine, but it only kicks the can one pace down the road. What we will need sooner or later is to show how the behaviour pattern in question arises from more general principles of surviving and reproducing in different kinds of environments: an *ultimate explanation*. For example, we might point out that people doing dangerous manual jobs or living in hostile environments tend to die anyway, for other reasons, before the age at which smoking starts to really kill you. Thus, the payoff for foregoing the pleasures of smoking may be less for them than for people living in under other conditions.[4] And we assume that people respond, sooner or later, to payoffs in the currencies of survival and reproduction. This kind of explanation has a few things going for it: it's non-obvious; it uses information not contained in the original observations; and it connects to broader expectations about evolved Darwinian creatures, such as that they should suit their behavioural strategies to the ecological circumstances they experience.

This ultimate explaining is a good thing, but we do tend to be rather smug about it. Look at you lot, rearranging your proximate deckchairs on the deck, we seem to imply, whilst we alone are looking beyond

---

4   See The mill that grinds young people old, this volume.

the prow, to the causally primal iceberg looming out of the sea. This book is about honest self-assessment, though, and in this spirit, I have to make the following admission about us evolutionary folk: the ultimate explanations we proffer are sometimes not as ultimate as we make them out to be. They also suffer from a lot of the same vagueness and indeterminacy as the more proximate frameworks we like to claim we are going beyond. It's healthy to admit this. And it's also healthy to understand that in science, with the possible exception of theoretical physics of the most fundamental kind, it's always the case that your explanations will themselves require deeper explanations in their turn (yes, even for us evolutionary folk). One person's *explanans* always ends up being someone else's *explanandum*. It's a food web of indefinite size, stretching off in every direction.

§

This whole business of explanation is very much tied up with having something called a theory. The function of theories is sometimes said to be able to predict future cases, at least up to relative statistical likelihoods. There is some truth in this, but at least as important a role for theories is to shed light on *why* things happen. For example, say I passed a vast archive of historical social and economic variables, and the results of elections from the same countries in the same years, through a machine-learning algorithm, to try to find regular relationships. Afterwards, I find the algorithm can predict election results in novel cases with 75% accuracy. Would I then have a *theory* of electoral outcomes? Not, it seems to me, without a lot more work. I would have to show which variables the machine-learning algorithm had given most weight to, and then relate these to some kind of general conception of humans as decision-makers: what they like, what they don't like, when they stick, when they shift. The aggregate behaviour of the electorate might not read off from thinking about a single representative voter; different sectors of the electorate might have different experiences and might respond to them in different ways, and there could be complex social dynamics at play. My theory might need to take this into account. Nonetheless, to have a theory, I would need not just to gain predictive statistical power over election outcomes, but also to gain epistemic power: the ability to state in comprehensible terms why elections turn out as they do, using some generalizations about voters, their voting, and their interactions.

Theories are devices that come in diverse kinds. The kind of theory lay people imagine scientists having is what I will call the *Newtonian*. The properties of a Newtonian theory, in my sense, are as follows. Only minimal and general properties of the situation are needed as inputs for the theory to do its work. If you are going to fire a cannonball into the air, on earth, then I can tell you that if you fire it off on flat ground at 45° at an initial velocity of 100 metres per second, it will travel about 1,020 metres and reach a maximum height of 255 metres at the mid-point of its flight. It will do this because it will be decelerated in the vertical dimension of its motion at a known constant rate due to earth's gravity. It doesn't matter whether the ball is black or pink or bears the colours of West Allotment Celtic; whether you do it on a Wednesday, under Scandinavian egalitarianism, or in anger at being spurned by your lover.

The theory has no real wiggle room from person to person. If two scholars apply Newtonian mechanics to the same problem, they must both conclude with the same predictions. If they don't, at least one of them has simply made a mistake. It should be relatively straightforward to look at the working and see where this has occurred. There is no: 'she's a Newtonian, but she brings a more modern sensibility and a command of the African evidence to the picture, and so she concludes the ball will fly 1,023 metres rather than 1,020'. The theory is a somewhat stable historical object. The Newtonian mechanics of today is just the same as the Newtonian mechanics of 100 years ago, and makes exactly the same predictions (there is only one possible prediction for a given set of inputs). We learn more about the world over time, for example that the theory doesn't do such a good job for things that are very small or moving very fast, but the theory itself is a fairly definite and stably identifiable entity.

Not all things that get called theories have the Newtonian properties. Take, for example, 'social practice theory'.[5] As I understand it, this theory presents an alternative both to rational actor models, which see people as free decision-makers with inherent preferences that they seek

---

5   See Schatzki, T. (1996). *Social Practices* (Cambridge: Cambridge University Press); or Reckwitz, A. (2002). Toward a theory of social practices: A development in culturalist theorizing. *European Journal of Social Theory* 5: 243–63, https://doi.org/10.1177/13684310222225432. These two texts have rather different things in mind. This reinforces the point I am making, since I have seen them both cited as descriptions of what social practice theory, or the theory of social practices, contends.

to satisfy, and acculturation models, which say that people do what their culture or society conditions them to do. The basic premise of social practice theory is that people will do what seems best to them, but are not asocial, cultureless, pastless, bodiless, and omniscient decision-making demons as implied by some micro-economic models. Instead, what seems best to them is limited by the habits, rules, norms, and understandings that they have absorbed through their daily lives in their social environments, that they also play an unwitting role in perpetuating. They are agents, but agents situated within a particular local field of social practices, a field that cannot be stood outside, or reinvented from scratch.

I'll call theories like this *recipe-for-a-recipe* theories. Their properties are a mirror-image of the Newtonian ones. Let's say we want to understand how people will respond to a social change, such as petrol being made 10% more expensive due to concerns about carbon emissions and climate change. Social practice theory does not make a simple general prediction, like 'car use will fall by 17%'. Instead, it says, we would need to know a lot of things about the context. How is car use embedded in people's daily practices; what social rules are there; what are normative pressures on them; what practical knowledge do they have of alternative modes of transportation; and so forth. The theory does not give us a prescriptive recipe for cooking up a prediction, as Newtonian mechanics did. Instead, it points us toward a flexible but not completely open-ended list of ingredients we ought to seek more information on in order to begin studying the problem, and hence make an appropriate recipe to then cook up a prediction (or more likely, retrodiction). The theory does not uniquely pre-specify what the relative proportions of these ingredients will be for the present case, nor how they will interact. It follows that two scholars can both employ social practice theory competently and without error, and yet come up with very different expectations, not just for two slightly different cases, but even for the very same case. There can be different emphases within the broad envelope of the theory, and the theory itself will drift over time, with different elements becoming more or less central.

The corollary of the comforting elasticity of recipe-for-a-recipe theories is that it is quite hard to say that they are wrong. If the cannonball in our previous example doesn't fly in a parabola and go 1020 metres plus or minus a few centimetres, then Newtonian mechanics gets chucked out as a theoretical framework for projectiles on earth. Discipline for social practice theory is more nebulous: almost any pattern of findings, *ex post*, can be parsed in a way that is compatible with the theory. The theory itself can update in the light of new evidence and new priorities; or it can fall from use in favour of some other recipe for making recipes that, like a new musical style, seems more interesting to the current generation. It might at best be shown to be more or less useful; there is almost no observation I can think of that would inflict it a critical blow.

There's a third type of theory I would like to mention, and that's the *inductive*. An example is the 'purse versus wallet' theory.[6] This theory says that increasing household income through giving it to mothers has greater positive effects on childrens' outcomes than via giving it to fathers, because of different ways the two genders spend their money. Unlike a recipe-for-a-recipe theory, it's pretty clear what this theory predicts for a given case within its domain (maybe not the size of the effect, but certainly its direction). Unlike Newtonian mechanics though, the main grounds for this theory seem to be largely, 'we have looked at some previous cases, and that's how it often worked out before', rather than any more general principles. (A quibble: Newton had undoubtedly looked at some previous projectiles and seen that they flew in parabolas prior to coming up with his theory; and perhaps you could found the 'wallet versus purse' theory on some more general first principles to do with the two sexes and evolutionary fitness. Nonetheless, the two cases do feel rather different.) So now we have defined three species of theory: Newtonian; recipe-for-a-recipe; and inductive. It's time to ask: when we construct evolutionary theories of human behaviour, which species of theory are we constructing?

§

When we make the evolutionary gambit, we sincerely feel Newtonian. It feels as if by beginning our Introduction 'Evolutionary theory predicts...', we have connected our claims to the might of biological science, and specifically to the considerable epistemic and formal power

---

6   Discussed by Cooper, K. and K. Stewart. (2013). Does money affect childrens' outcomes? A systematic review. *The Joseph Rowntree Foundation*. Downloaded from: https://www.jrf.org.uk/sites/default/files/jrf/migrated/files/money-children-outcomes-full.pdf

of Darwinian algorithms. But in reality, lots of 'evolutionary' theories about human behaviour are far from Newtonian, and don't have such solid explanatory or formal support as you might think. They are really just recipes for recipes, or patterns discovered by induction, and a good theoretical biologist would still want to ask 'yes, but under what conditions would *that* evolve?'. In this regard, we are not so different from any other kind of social scientist. I have a case study that illustrates this very clearly: the recent enthusiasm for 'life history theory' as an explanation for diverse human behavioural phenomena from risk-taking to obesity, schizophrenia to savings. If you want to see the kind of research I am talking about, just type 'life-history theory' or 'life-history strategies' into your literature search engine of choice and follow up the recent human-focused references.

When human behavioural scientists invoke 'life history theory' as an explanatory framework, there are a number of related things they might actually be up to. I will call three prominent ones 'enterprise 1', 'enterprise 2', and 'enterprise 3'. Not all work on 'life-history theory' falls into any of these enterprises. Moreover, I have no objection to any of them—indeed, have contributed to some—but the role being played by the term 'theory' within them does bear some examination.

Enterprise 1: often, what researchers invoking life-history theory are doing is asking whether the behaviour under study covaries with a number of other traits, particularly those to do with the timing of reproduction (for example, age at first menarche, age at first sexual intercourse, or age at first childbearing). The idea here is that human psychological and reproductive traits, rather than each varying independently, covary along a principal axis, the 'fast-slow continuum'.[7] At the fast end, we have early maturation and childbearing, along with which allegedly go a high rate of future discounting, proneness to violence and coercion, impatience, obesity, certain moral and social attitudes, and even certain psychiatric disorders. At the slow end, we have late childbearing, high parental investment, and all the opposite

---

7 For a reviews and critiques of enterprise 1, see: Copping, L. T., A. Campbell and S. Muncer. (2014). Psychometrics and life history strategy: the structure and validity of the High K Strategy Scale. *Evolutionary Psychology* 12: 200–22, https://doi.org/10.1177/147470491401200115; and Gruijters, S. and B. Fleuren. (2018). Measuring the unmeasurable: The psychometrics of life history strategy. *Human Nature* 29: 33–44, https://doi.org/10.1007/s12110-017-9307-x

behavioural and psychological traits. Sometimes the mere demonstration that different traits covary along a principal axis is presented as if this were an explanation of these traits, and also a confirmation of the utility of life-history theory. There are, however, lots of reasons things might be correlated with one another, and demonstrating a correlation is very different from explaining it.

Enterprise 2: Sometimes scholars invoking life-history theory are doing more than just establishing covariation between traits. They are also trying to demonstrate that those traits relate to the ecology in which people live. In particular, they may be testing whether 'fast' behaviours are differentially likely to occur in places where life prospects are poor or uncertain.[8] This relates to an intuitively appealing argument that if the environment is harsh (for example, uncontrollable mortality is high), then you need to get on with life quickly and at least get some reproduction done while you can, whereas if the environment is benign you can take longer and invest more in temporally distant outcomes. Note that enterprise 2 is in principle independent of enterprise 1: it could be that traits covary along a principal axis, but for some other, completely different kind of reason than the one argued in enterprise 2.

Enterprise 3: If 'fast' behaviours occur particularly in 'harsh' environments, there are a number of ways this could come about, from the slow march of genetic selection at one extreme to rational real-time decision-making at the other extreme. Enterprise 3 concerns the particular claim that experiences in the first few years of childhood are particularly important in setting how 'fast' or 'slow' a person turns out later on.[9] The idea here is that people are born not knowing what their adult environment is like, but that things like the stability of the family, how their parents behave, and so forth, serve as cues that over evolutionary time have carried useful information about their adult worlds. Thus, natural selection has favoured mechanisms that

---

8   See for example: Nettle, D. (2010). Dying young and living fast: Variation in life history across English neighborhoods. *Behavioral Ecology* 21: 387–95, https://doi.org/10.1093/beheco/arp202

9   See for example: Brumbach, B. H. et al. (2009). Effects of harsh and unpredictable environments in adolescence on development of life history strategies: A longitudinal test of an evolutionary model. *Human Nature* 20: 25–51, https://doi.org/10.1007/s12110-009-9059-3

effectively say 'if this crazy stuff is going on in childhood, I need to get ready for a world where I am going to need to be a fast adult'.

There's actually a lot of evidence compatible with the idea that childhood adversity affects reproductive behaviour and many other adult outcomes besides: with that I have no quibble. I just want to point out that enterprise 3 is not deducible from the other two enterprises. You could believe that there is a fast-slow continuum, but that it is not related to environmental harshness; that there is a fast-slow continuum related to environmental harshness but childhood experiences do not serve as cues to speed you up or slow you down; or that there is a fast-slow continuum and childhood experiences move you along it, but for reasons that have nothing to do with those experiences being evolutionarily valid cues to prevailing environmental harshness. So which of these various enterprises is the core claim of 'life-history theory'; and, more importantly, which of the various enterprises has its explanatory basis in evolutionary theory?

§

There is an area of evolutionary biological theory called 'life-history theory'. In fact, it is not any single theory, but a body of mathematical methods for making theories, theories about how natural selection would shape patterns of growth, reproduction and ageing under different ecological circumstances.[10] These methods have been applied to many different scenarios, and the general conclusion seems to be: all kinds of different things can evolve, depending on the details of the ecology and demography. And that seems to be borne out in nature: we see everything from salmon that go out in a single blaze of reproductive glory, to puffins that do a little bit of reproduction year after year for ages. There certainly are life-history models that show that if the risk of uncontrollable mortality is high, one should expect early reproduction to evolve, even at the expense of growth or self-repair.[11] This prediction depends on a lot of things, though: small tweaks in assumptions about,

---

10 Stearns, S. C. (1992). *The Evolution of Life Histories* (New York: Oxford University Press) is a classic text, and possibly more widely cited than read.

11 For example, Cichoń, M. (1997). Evolution of longevity through optimal resource allocation. *Proceedings of the Royal Society*: B 264: 1383–8, https://doi.org/10.1098/rspb.1997.0192

for example, what limits population growth, and when in the life cycle mortality acts, can lead to the prediction that higher mortality will delay reproduction, or have no effect.[12]

Because of this, it is not really correct to say 'life-history theory predicts X...'. Really what one ought to say is 'this particular life-history model, using this particular set of assumptions, predicts X...'. Then as well as testing prediction X, you would also want to establish that the assumptions were appropriate for the system you were working on. Now you might argue in the following way: in practice, we know that animals facing higher mortality regimes often evolve earlier reproduction. We know this not just from correlational evidence, but even from experimental evolution.[13] So the best class of theoretical models is probably the class that correctly recovers this phenomenon. I have some sympathy with this argument, but note that theory and evidence have changed places. Rather than life-history theory predicting *a priori* that this phenomenon will occur, we see that phenomenon often does occur, and then use that discovery to fix the theory. So it is not so much a case of 'life-history theory predicts that environment harshness will lead to the evolution of earlier reproduction...' as 'in practice, environmental harshness often leads to the evolution of earlier reproduction, and this motivates us to search for selective reasons why that might be true'. If you say 'life-history theory predicts that environment harshness will lead to the evolution of earlier reproduction...', as we often do in enterprise 2, then you are using the word 'theory' in an inductive, not a Newtonian sense.

What about enterprise 1, the idea that multiple different behaviours covary along a 'fast-slow' principal axis? Authors in enterprise 1 are

---

12   This specificity was given vigorous restatement recently by Baldini, R. (2015). Harsh environments and 'fast' human life histories: What does the theory say? *BiorXiv*, https://doi.org/10.1101/014647

13   Correlational evidence: Promislow, D. E. L. and P. H. Harvey. (1990). Living fast and dying young: A comparative analysis of life-history variation among mammals. *Journal of Zoology* 220: 417–37, https://doi.org/10.1111/j.1469-7998.1990.tb04316.x. Experimental evolution: Reznick, D. A. et al. (1990). Experimentally induced life-history evolution in a natural population. *Nature* 346: 357–9, https://doi.org/10.1038/346357a0. What the experimental work actually shows is that greater predation risk only leads to the evolution of earlier reproduction if the predation acts in the adult, not the juvenile, stage of life. This supports the main conclusion of the theoretical models—that what evolves depends on the details of the demography.

unanimous in expressing the idea that the existence of such a principal axis is a basic prediction of life-history theory. Kimberley Mathot and Willem Frankenhuis recently conducted a systematic review of relevant models and concluded, perhaps surprisingly, that 'there is, at present, little formal theory' relating to the reasons why a single fast-slow principal axis would evolve.[14] If there is 'little formal theory' on the question, one has to ask, why do so many people believe the existence of such an axis to be a basic prediction of life-history theory?

The origins of the idea of the fast-slow continuum are in fact empirical more than theoretical. If you get empirical data on different species, for variables like age at first reproduction, litter size, duration of gestation, and duration of lactation, and stick them into a big correlation matrix, then *empirically* you discover that there is a principal axis, with late, slow and long species at one end, and early, fast and short species at the other.[15] So we can only really say that the axis is predicted by theory if by theory, we mean induction. What is really an empirical regularity has somehow morphed into being widely considered a theory. But importantly, in the original empirical analyses, the unit was the species, not the individual, and all the traits entered into the analysis were reproductive ones. The idea that you would get a single axis when comparing different individuals of the same species, and in particular that non-reproductive behavioural traits would also fall along this same axis as reproductive ones, is certainly out there in the literature of non-human biology, but not clearly supported by current evidence.[16]

It follows that when we do enterprise 1 in humans, we are not testing a prediction stemming directly from formal evolutionary theory in some Newtonian manner. We are taking an empirical pattern of co-variation seen across species, and arguably perhaps within some non-human species, and then looking for something vaguely analogous in human behavioural variability. I don't think there is necessarily anything wrong

14   Mathot, K. J., and W. E. Frankenhuis. (2018). Models of pace-of-life syndromes (POLS): a systematic review. *Behavioral Ecology and Sociobiology* 72: 41, p. 11, https://doi.org/10.1007/s00265-018-2459-9

15   Promislow, D. E.L. and P. H. Harvey. (1990). Living fast and dying young: A comparative analysis of life-history variation among mammals. *Journal of Zoology* 220: 417–37, https://doi.org/10.1111/j.1469-7998.1990.tb04316.x

16   See Royauté, R. et al. (2018). Paceless life? A meta-analysis of the pace-of-life syndrome hypothesis. *Behavioral Ecology and Sociobiology*, 72: 64, https://doi.org/10.1007/s00265-018-2472-z

with doing this, but one has to wonder in what sense we are 'using life-history theory' or 'testing the predictions of life-history theory'. And indeed, one has to wonder exactly what the theoretical entity is that is subject to potential falsification here. Let's say we do an empirical study of a whole set of psychological traits within a human population and find that they don't really vary along a single axis. What exactly would be the endeavour whose credibility is undermined? Evolutionary life-history theory? Its application to humans? The generalization from reproductive traits to these particular psychological traits? It is not clear.

§

So far I seem to have argued that 'life-history theory' as it gets used in application to human behavioural traits is really a kind of extension of an inductive regularity, rather than a Newtonian theory. Actually, it's not even always that: it's sometimes a recipe for a recipe. My exemplar here is interesting experiments showing that people from different childhood backgrounds seem to respond very differently, in terms of their behavioural intentions for the future, to imagined scenarios evoking a world of harshness and scarcity.[17] So far so good, but these experiments are explicitly framed within 'life-history theory'. Thus the implication is that life-history theory either predicted these different responses *a priori*, or at least provides some major explanatory insight into them.

Life-history theory here is clearly being interpreted in terms of enterprise 3: there are fast and slow ways of behaving, and your childhood affects where you are on the continuum. Fine. But the experiments have two independent variables: childhood experience, and the imagined scenarios (either harsh world, or control). All enterprise 3 says here is 'somehow childhood experience will turn out to matter'. Equally compatible with the general contention of enterprise 3 would be: *only* childhood experience, and not the content of the current scenario, affects behavioural intentions; childhood experience and the current scenario

---

17   Griskevicius, V. et al. (2011). The influence of mortality and socioeconomic status on risk and delayed rewards: A life history theory approach. *Journal of Personality and Social Psychology* 100: 1015–26, https://doi.org/10.1037/a0022403; Griskevicius, V. et al. (2011). Environmental contingency in life history strategies: The influence of mortality and socioeconomic status on reproductive timing. *Journal of Personality and Social Psychology* 100: 241–54, https://doi.org/10.1037/a0021082

both matter, and their effects are additive; or they matter synergistically in some kind of way (adverse childhood experience deadens you to cues of current harshness, or childhood experience sensitizes you to cues of current harshness). In short, 'life-history theory', as the phrase is being used, would be compatible with any conceivable pattern of results other than the one in which childhood experience does not matter at all. Thus, I have to ask: what sense of 'theory' is it, when 'life-history theory' as applied in this instance is compatible with most of the possible empirical outcomes?

I think the answer is that 'life-history theory' is being used as a recipe for a recipe. It denotes the general expectation that behaviours can be thought of as concerning doing things soon and fast, or later and slowly, and that one's childhood experience will make some kind of difference to one's propensities along this continuum. Exactly what kind of difference, and how childhood experience will combine with other situational factors, is a matter for further determination. The theory does not say. This is fine, I suppose, but we need to take away two things. First, this kind of theorising has only the vaguest of connections to the formal body of life-history models constructed by evolutionary biologists. And second, 'life-history theory' as used here is no more Newtonian than any other social-science theory. In fact, it looks almost exactly like the theory of life-course epidemiology, which I have written about elsewhere.[18] Life-course epidemiology basically says, all the things that happen to you over the course of life, including in particular childhood, are going to affect patterns of health and disease. How the different influences combine (additively, non-additively, etc.) is subject to further determination; indeed, the theory itself will be narrowed down in this regard according to what we find out empirically.

§

What lessons do I take away? First, narrowly, to people like me who want to apply evolutionary ideas to human behaviour. We shouldn't be so sloppy as to say 'Evolutionary theory predicts X...', or 'life-history theory predicts X...' Evolutionary theory can predict a lot of things.

---

18 Nettle, D. and M. Bateson. (2017). Childhood and adult socioeconomic position interact to predict health in mid life in a cohort of British women. *PeerJ* 5: e3528, https://doi.org/10.7717/peerj.3528

Instead, if what we mean is 'the evolutionary model by Bloggs (2018) makes the prediction, that, under assumptions 1, 2 and 3, X will happen', then we should say exactly this. If what we mean is 'Empirical findings from non-human animals suggest that X often happens', then we should say that and cite the evidence. And if what we mean is 'I happened on some patterns that I want to endow with the gravitas and authority of the most famous and respected meta-theory in the life sciences because it makes them sound better', then, well, that's just bad.

Another take-home is that we should actually do more life-history modelling, indeed more mathematical modelling in general, to try to provide stronger theoretical underpinnings to the observations we make. Mathematical models are only heuristic devices, and they don't solve all your scientific (or even theoretical) problems. They are very useful though, as instantiations of what your theory really is: formalising makes a theory into a stable, reproducible entity that can easily be queried. Models are useful in forward-engineering from starting assumptions to predictions, because verbal arguments are notoriously ambiguous, and informal intuitions about what would follow from what are often just wrong. Having a mathematical model is a way of showing rigorously that if I make this particular set of assumptions, then the prediction to which I am led is exactly the following. Rather than: once I got the data (or someone else gathered some data) I realised that quite possibly the predictions of my theory were not necessarily the ones I originally thought they were. And models are useful in reverse-engineering, from phenomena back to explanation. Say aggression is correlated with the risk of predation. Maybe there is some adaptive reason these two things get coupled. Now, under what ecological and demographic assumptions could such a coupling emerge? Then you can start to ask whether those assumptions seem plausible for ancestral humans (or whatever system you are studying).

I should note, however, that even if we made such models, even if those models made predictions, even if we tested those predictions, and those predictions were supported, it still would not be (complete) explanation yet. The classic models of life-history theory, indeed of behavioural ecology in general, are mostly only approximations. That's because they have no explicit population genetics, or only something very simplified. Yet what would actually evolve, presumably, would be alternate forms of genes in populations, populations where many genes were interacting and traits had a complex genetic basis. So even when we are done with making optimality models that explain to our satisfaction why men are more aggressive than women, or people in harsh environments reproduce at a young age, a whole other group of people, theoretical evolutionary geneticists, would see that just as a heuristic starting point, a sketch of a proposal still to be properly explained. And they would need their own theories and models to do their bit. Explanation is never done: it's just passed along the row. It reminds me of a conversation I once heard between two physicists. One said that he had been able to prove mathematically that some effect should occur under some set of circumstances. 'I mean prove to the satisfaction of a physicist, of course', he added. Proof to the satisfaction of a mathematician was a completely different issue.

A similar point can be made about providing a mechanism. People often say to us behavioural ecologists, yes, that's an interesting evolutionary theory, but what's the mechanism by which it would be delivered? I was studying early-life adversity and ageing in starlings.[19] I was pleased with myself because we were measuring oxidative stress, a possible ageing factor, at the cellular level. I explained my plans to some cellular biologists. Interesting, they said. If you did find a difference in oxidative stress, what do you think the mechanism would be? What? Oxidative stress was, for me, the mechanism. Indeed, it was about the most mechanistic I had ever got. For them, my oxidative stress measure was just some crude phenotypic summary. How the oxidative damage to lipids picked up by the assay actually came about was a hole where a mechanism needed to be placed. Will it never end? Probably not. Just as one person's *explanans* is another person's *explanandum*, one person's mechanism is clearly another person's black box.

More broadly, it's clear that when researchers use the term 'theory', they are not referring to a homogenous class of entities. It would perhaps be helpful to use more precise terminology to refer, respectively, to specific hypotheses, inductive generalizations, mathematical models, recipes, recipes for recipes, and so on. Given the indefinitely large

---

19   Nettle, D. et al. (2015). An experimental demonstration that early-life competitive disadvantage accelerates telomere loss. Proceedings of the Royal Society B: *Biological Sciences* 282: 20141610, https://doi.org/10.1098/rspb.2014.1610

food-web in which we all operate, it would also be quite handy if theories, like buses, had an origin and destination clearly displayed on them. The theory of planned behaviour, for example, is a well-formed inductive theory whose destination is behavioural decisions, and whose origin is immediate psychological factors like beliefs and intentions. It never claimed to serve earlier stations on the line, and should not therefore be criticized for not doing so. A theory might usefully say: I'll pick you up at known regularities of individual human cognition, and drop you off at cross-cultural regularities in the content of literary stories. If you want to get on any earlier (e.g. where do the known regularities in individual cognition come from?), you will need to take an additional bus.

This picture casts an unflattering light on the idea, sometimes raised with rather messianic zeal, that the human sciences might one day be unified under a single grand theory. That idea is like saying that every bus stop in the city should be served by the same bus: hardly a recipe for getting from A to B any time soon. Surely the unification we should be looking for is not that a single bus goes everywhere, but rather that, using a network of buses that is reasonably well integrated, it is eventually possible to get from any starting point to any destination, using several types of theory along the way. The vision is the one beautifully expressed by Melvin Konner in the preface to *The Tangled Wing*:[20]

> A good textbook of human behavioral biology, which we will not have for another fifty years, will look not like Euclid's geometry—a magnificent edifice of proven propositions deriving from a set of simple assumptions—but more like a textbook of physiology or geology, each solution grounded in a separate body of facts and approached with a quiverful of different theories, with all the solutions connected in a great complex web.

And by the way, in closing, this for me is where we can shed a bit more light on the special explanatory status of evolution for the life sciences, and therefore for their subset the social sciences. We all know Dobzhansky's famous dictum 'Nothing in biology makes sense except in

---

20   Konner, M. (2003). *The Tangled Wing: Biological Constraints on the Human Spirit* (2nd ed., New York: Holt, p. xv). I am grateful to Karthik Panchanathan for introducing me to this quote.

the light of evolution'.[21] Continuing with my affordable public transport analogy, we could read this dictum in two ways. The strongest reading is the requirement that every bus, whatever its destination, must have evolution as its origin point. Even I, an enthusiast, can see that this is much too strong a requirement to be sensible. The second reading gives evolution a weaker, but still rather special, status. If you rode each bus back to its point of origin, and there picked up another bus and rode that one back to its origin, and so on and on, then wherever you started out, evolution would sooner or later, in one way or another, be the place you ended up. You might visit a lot of different and exotic locations along the way, but you can't really avoid getting back to evolution at some point, because we are embodied creatures who arose through a historical process that also produced the other organisms with which we share the earth. 'What we are supplying', as Ludwig Wittgenstein put it, amount in the end to 'remarks on the natural history of human beings'.[22]

---

21 Dobzhansky, T. (1973). Nothing in biology makes sense except in the light of evolution. *American Biology Teacher* 35: 125–9, https://doi.org/10.2307/4444260

22 Wittgenstein, L. (1953). *Philosophical Investigations* (Oxford: Blackwell, section 415).

PART TWO

6. The mill that grinds young people old

One and the same cause
Wears out our bodies and our clothes

–Bertolt Brecht, *A worker's speech to a doctor*

I am in the cemetery again. It's a good place to meditate on the conditions of life, and on the relationships between biology and social science. This is the cemetery of 'neighbourhood B', one of the fieldwork sites of my ethnographic project *Tyneside Neighbourhoods*.[1] That project was about life. Here I am reflecting on ageing and death.

Neighbourhood B is in Newcastle upon Tyne. Walking distance from two huge universities, from a large teaching hospital, a cycle ride from offices of regional and national government, it is nonetheless one of the most deprived places in Britain. People working in dangerous heavy industries lived and died here. Later, the neighbourhood atrophied along with the industries that bore it. Its population declined and its future became unclear. It is not a bad place. It persists quietly because there is nothing else it can do, persists despite stagnation and economic precariousness for its residents, squalor in its structures, and ever-greater retrenchment of its public services. The cemetery is an odd mixture of municipal decline and gaudy activity. The fine mausoleum and other elaborate Victorian buildings are all falling down. Maintenance is limited to some movable iron fencing panels to keep people out. Pigeons roost in collapsing roofs. The older headstones have fallen over, or else been laid down before they injure someone. But many of the younger headstones—black marble and gilt—are islands of activity in the

---

1   Nettle, D. (2015). *Tyneside Neighbourhoods: Deprivation, Social Life and Social Behaviour in One British City* (Cambridge: Open Book Publishers), https://www.openbookpublishers.com/product/398/, https://doi.org/10.11647/obp.0084

https://doi.org/10.11647/OBP.0155.06

grass. Fading children's bears, balloons, photos, withering flowers, hand-written messages—some of these on the graves of people who died twenty or even thirty years ago.

The place is written through with human biology, for what could possibly be more biological than the cessation of our organismality, the cessation of all of those metabolic and physiological processes that make us someone rather than an inanimate object? Yet this is also a place rich in social meaning and social pattern. This is not an accidental coupling, some unlikely final juxtaposition of two worlds—the 'social' and the 'biological'—that in life flow separately, have little fundamentally to do with one another. This place derives its social meaning from the very fact that death (and therefore life) is a biological process. Its social meaning is incomprehensible if not grounded in that fact. That is what brings the mourners with their balloons and their bears, the undeniable and impassive *biology* of the situation. But though these deaths were biological events, their determinants (or the determinants of their immediate determinants) belong very clearly in the domain of the social. Here in the cemetery, there is no space between the biological and the social. The graves mark their indissoluble unity, which was as true in life as it was in death.

§

Wandering around, I am struck how many of these people died young. Here's Susanna E., died at 44; Gemma G., aged 10; Jay R., clearly a big Newcastle United supporter to judge from his headstone, aged 19. Here's Paul C., dead at 33, next to his dad James, dead a couple of years earlier at age 55. There are several graves of babies. Life round here seems, if not nasty and brutish, then certainly short. Of course, I could be guilty of confirmation bias, of noticing the stand-out young ones. So I get interested and start to collect some data.

I complete a survey of the graves dating from 1990 onwards, noting sex and age at death. It is not perfect, since I get lost crossing and re-crossing the cemetery, and some more recent deaths are put into older established family graves, confusing the inclusion criteria.

Still, I end up with a sample of nearly 200 deaths. With this sample I can estimate the probability density function for age at death. What this function tells you is not the probability of dying at any particular age (you would need data from the living for that). What it tells you is, given that you died, the probability of being any particular age at the time. This is a relevant consideration for life; we've all got to die at some age, and it would be nice to have a sense of the distribution of likelihoods for what that age will be.

The resulting density is shown in figure 1. It looks different for men and women, which we ought to expect. The average age at death is about 61 for men, and nearly 69 for women. 61 years. The current (2014) life expectancy at birth for Afghanistan is 60.[2] More important than the low average is the variability in ages at death. As a man from round here, it is most likely that my age at death will be in the 60s, but it could be any age, from the day of my birth onwards. As you can see from figure 1, the likelihood of death coming in my twenties is really not negligible. For the women, late seventies is the most likely time, but again, it could well be earlier (though not as likely to be under 40 as is the case for the men), just as it could be later. A statistic that expresses this imprecision in when we are going to die is the standard deviation of age at death. The mean of 61 says that the average age at death of all the men is 61, but the standard deviation of age at death of almost 20 years says that a typical individual's age at death is higher or lower than the average by two decades. So you could be 41, or 81, without being in any sense exceptional.

This is all very atmospheric, you might say, but not very scientific. This is one cemetery with no comparison data. Not everyone who dies is buried here; maybe there are biases towards memorializing those who died young. Quite right of course, but the cemetery is just an illustration of something we know to be true from much more systematic national data. Poor people die relatively young in contemporary Britain. The size of the disparity depends a little how you do the calculation. If you do it by individual social class, it is probably around 6 years for men and

---

2   Information from http://data.worldbank.org/indicator/SP.DYN.LE00.IN

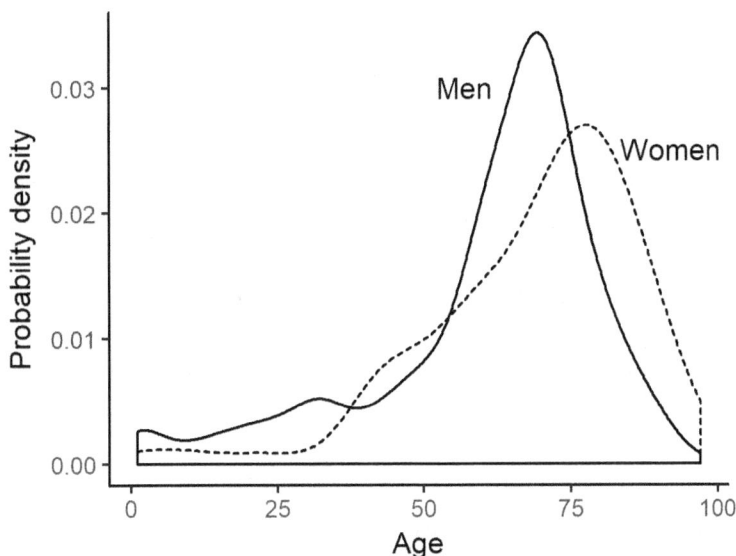

Figure 1. Estimated probability density functions for the male and female ages at death in Neighbourhood B cemetery, graves established since 1990, survey of January 23rd 2017.

5 years for women.[3] If you do it by place, comparing not individuals by their occupation, but communities by their overall levels of deprivation and want, the disparities are more like 8 and 6 years.[4]

§

We first meet Paris, in Dickens's *A Tale of Two Cities*, in an extraordinary passage at the beginning of Chapter 5. In the neighbourhood of St. Antoine, a cask of wine has been dropped and broken in the street. Suddenly, there are the inhabitants, scooping wine with their hands from between the cobble stones; making dams out of mud to drink the resulting pool; mopping wine up with handkerchiefs to squeeze into the mouths of their infants; even champing on wine-rotted fragments of

---

3   Trend in life expectancy at birth and at age 65 by socio-economic position based on the National Statistics Socio-economic Classification, England and Wales: 1982–1986 to 2007–2011. Office for National Statistics statistical bulletin released 21st October 2015. Downloadable from www.ons.gov.uk.

4   Inequality in Health and Life Expectancies within Upper Tier Local Authorities: 2009 to 2013. *Office for National Statistics statistical bulletin* released 20th November 2015. Downloadable from www.ons.gov.uk.

barrel. They leave whatever they were doing, the painful drudgery of their daily lives, in mid-action, for a transitory wine frolic, an expected carnival.

The people of St. Antoine, Dickens tells us, had undergone a terrible grinding and re-grinding in the mill of poverty. But what kind of mill is the mill of poverty? Dickens: 'The mill which had worked them down, was the mill that grinds young people old'. The mill of poverty is a mill of ageing: the effect of poverty is to age you faster. This is an important idea, not least because it suggests how an intuitively 'internal' and 'biological' process, ageing, is profoundly affected by the indubitably 'social' processes of deprivation and inequality happening beyond the body envelope.

§

I want to take Dickens' analysis of St. Antoine seriously, in two ways: first, the effects of poverty should be considered as ageing; and second, that the spontaneous frolic with the windfall of wine—the behaviour of the residents of St. Antoine—is somehow systematically connected to the ageing effects of poverty.

What evidence could we adduce in support of the idea that the effects of poverty constitute ageing? We need to define what ageing is. Biologists define ageing as the deterioration of an individual's biological performance over his or her life. Influential current theories of ageing suggest that this decline is due to the accumulation, in the body, of unrepaired damage.[5] DNA becomes oxidated and unreadable; key populations of stem cells lose capacity; mechanisms lose their shape and capacity to rebound; all due to the continuous assaults of chemistry and physics upon our bodies. There is no internal ticking clock; organisms are not programmed to self-destruct after some particular delay. For this reason, the pace of ageing can vary wildly from individual to individual.[6] It is not time *per se* that it is doing the work: it is the net

---

5   Kirkwood, T. and S. Austad. (2000). Why do we age? *Nature* 408: 233–8, https://doi. org/10.1038/35041682

6   Belsky, D. W. et al. (2015). Quantification of biological aging in young adults. *Proceedings of the National Academy of Sciences* 112: E4104–10, https://doi.org/10.1073/pnas.1506264112

effect of the rate at which damage accumulates, and the activities of the body to repair it.

So what evidence is there that poverty can accelerate ageing? Well, there is no greater decline in biological performance than becoming dead, so in one sense, if death comes earlier for poor people (and we have already seen that it does) then ageing is by definition faster. But we can see it clearly in declines in performance short of death, what medics call morbidity, as well. Poor people are in worse health than rich people at all ages, but importantly their health goes downhill with time more rapidly through adulthood. We can distinguish statistically between expectancy of life—the number of years you can expect to be alive at prevailing rates of mortality—and expectancy of health, which is the number of years of good health you can expect to have at prevailing rates of morbidity and mortality. Whilst the life expectancies of poor English communities are 6–8 years less than those of rich ones, the health expectancies are nearly 17 years less, for both sexes.[7] In the most deprived English neighbourhoods, health expectancy is only about 50. In my neighbourhood B, fully one third of 18–65 year olds have a serious longstanding health problem.[8] Deprivation and the deterioration of physical health as the years go by are so closely associated that you could almost use the latter as a measure of the former.

It is not just medical symptoms that show an effect of poverty: it is also underlying physiological processes. There is increasing interest in the idea that we should measure people's biological age (that is, where they are in the inevitable arc of biological performance) rather than just their chronological age, which is a very poor approximation for what is really happening to them.[9] We can do so using suites of 'biomarkers', namely bodily measurements that have the properties of (a) changing on average with increasing chronological age; and (b) predicting time

---

7   Figures from Inequality in Health and Life Expectancies within Upper Tier Local Authorities: 2009 to 2013. Office for National Statistics statistical bulletin, released 20th November 2015. Downloadable from www.ons.gov.uk.

8   Nettle, D. (2015). *Tyneside Neighbourhoods* (Cambridge: Open Book Publishers, p. 115), https://www.openbookpublishers.com/product/398/, https://doi.org/10.11647/obp.0084

9   Levine, M. E. (2013). Modeling the rate of senescence: Can estimated biological age predict mortality more accurately than chronological age? *Journals of Gerontology*, A 68: 667–74, https://doi.org/10.1093/gerona/gls233

until death better than chronological age does. One set of such markers measure inflammation. Inflammation is part of our immune response to injury and infection, and such the capacity to mount an inflammatory response is an adaptive one. With age though, the background levels in our bodies of molecules involved in the general inflammatory response increase. The levels of these molecules predict future serious disease better than chronological age does. Thus, inflammation markers in the blood (C-reactive protein and interleukin-6 are the most widely measured) serve as markers of biological age.

Many studies have found that poor people show higher levels of these inflammation markers than rich people, and not just when they are old.[10] It's true in mid-life, decades before most people die, and it is even true in adolescence.[11] The study on adolescents examined what it is about the lives of the less privileged that best explained (in a statistical sense of explain) their greater inflammation. The answer is sobering: they experience less happiness.

§

'If I'd known I was going to live this long, I would have taken better care of myself' is one of those quips with something so satisfying about it that it ends up attributed to many different people. It makes a kind of sense: the extent to which we orient our behaviour toward the future depends on how likely that future is to ever come about. Indeed, this is one of the key principles of some evolutionary theories of ageing. The mouse that invests so much in repairing its DNA that its DNA would continue to replicate fine for 10 years has probably wasted its effort. In the wild, 90% of mice are gone within 1 year anyway, from predation

---

10 E.g. Gruenewald, T. L. et al. (2010). Association of socioeconomic status with inflammation markers in black and white men and women in the coronary artery risk development in young adults (CARDIA) study. *Social Science and Medicine* 69: 451–9, https://doi.org/10.1016/j.socscimed.2009.05.018; Koster, A. et al. (2006). Association of inflammatory markers with socioeconomic status. *Journals of Gerontology* A 61: 284–90, https://doi.org/10.1093/gerona/61.3.284; Nettle, D. (2014). What the future held: childhood psychosocial adversity is associated with health deterioration through adulthood in a cohort of British women. *Evolution and Human Behavior* 35: 519–25, https://doi.org/10.1016/j.evolhumbehav.2014.07.002

11 Chiang, J. et al. (2015). Socioeconomic status, daily affective and social experiences, and inflammation during adolescence. *Psychosomatic Medicine* 77: 256–66, https://doi.org/10.1097/psy.0000000000000160

or cold. So mice have evolved to spend no more on DNA maintenance than necessary; instead, what they are really astonishingly good at is making baby mice while the sun shines. Engineers apparently get it too. An urban legend has Henry T. Ford instructing his engineers to tour the scrapyards of American looking for parts of his cars that never wore out. They found that the king-pins of the scrapped cars invariably still had life in them. His response: make the king-pin less well.[12]

Now we turn back to the residents of Dickens' St. Antoine, with their impromptu carnival of the spilt wine. They are conforming to an established stereotype about poor people: they value immediate opportunity (dropping what they were doing to consume during the day) over preparing for the future (the sawing of the wood stands neglected). Gillian Pepper and I recently reviewed the evidence that this stereotype contains a germ of truth: people living in poverty in Western countries do favour the present relative to the future more than their affluent co-citizens, in a number of different ways.[13] This orientation to the present is underpinned by a kind of fatalism and a belief in the role of chance. For many commentators these attitudes, these 'poor choices', become something to condemn morally, or attempt worthily to educate away, a psychological failing of poor people that is the root cause of their poverty.

But there is another side from which you can look at this. Here in my neighbourhood B cemetery, I ask myself: why not? Say I am a member of the C. family, whose graves I have just been looking at. The dad got 55 years, the son got 33. How much effort would I choose to make in, say, saving for a pension. Neither even reached the statutory pension age. Would I have smoked? Well, nicotine is a stimulant, giving you a pleasant buzz, and the really bad consequence, lung cancer, doesn't really start to hit until after age 45.[14] Half of deaths from lung cancer

---

12   This story can be found in many places, but was probably introduced into the folk culture of biology by Humphrey, N. (1976). *The social function of intellect. In Growing Points in Ethology* (P. P. G. Bateson and R. A. Hinde eds., Cambridge: Cambridge University Press, p. 303–17).

13   The evidence is reviewed in Pepper, G. V. and D. Nettle. (2017). The behavioural constellation of deprivation: Causes and consequences. *Behavioral and Brain Sciences* 40: e314, https://doi.org/10.1017/s0140525x1600234x

14   See: http://www.cancerresearchuk.org/health-professional/cancer-statistics/ statistics-by-cancer-type/lung-cancer/mortality#heading-One

are in the over 75s. You see the point: a lot of the decisions that poor people make start to make a kind of sense. Gillian Pepper struggled to find a name for this 'making a kind of sense'. She didn't want to use the term 'adaptive', since this has a technical meaning in biological theory, a technical meaning that was close but not identical to what she meant. She didn't want to use the term 'rational', since this can mean a number of different things, and for many, connotes the result of very extensive conscious deliberation, which she did not want to imply. So she settled on 'contextually appropriate response'. Living with a bias towards the present is a contextually appropriate response to the reality of poverty. Gillian (and I) are agnostic about whether this response is extensively reasoned through, or more automatic and sub-conscious, or a bit of both.

The best worked-through case of contextual appropriateness is the age of childbearing. The really big difference between the rich and the poor in Britain is not in how many children they have, but in when they have them. On average, this differs by at least a decade between the richest and poorest districts.[15] We can see this very clearly in the cemetery. Here's the grave of Nora W., dead at 23 but already a mum; Maureen O., dead at 49 but already a nana (grandma); Tommy D., dead at 62 but already a great-grandfather. You have to get on with it to keep the generation time this short. Commentators are fond of morally chastising the poor for their reproductive decisions, and laying all kinds of social ills at the door of early childbearing.[16] This is quite unjustifiable: the extremely late reproduction of middle-class people causes far more by way of medical problems and costs.

Instead, let us put the problem the other way around. It's quite a widespread human desire to hold one's grandchildren, to care for them whilst one is still hale and living. Those very commentators who lambast teenage mothers would probably endorse this aspiration whole-heartedly. In a paper a few years ago, I entered into the following thought experiment: Say I was a young woman and wanted to be able to expect, assuming my life and that of my daughter followed the average

---

15   Nettle, D. (2010). Dying young and living fast: variation in life history across English neighborhoods. *Behavioral Ecology* 21: 387–95, https://doi.org/10.1093/beheco/arp202

16   In the UK, early childbearing was a media and public policy obsession for a while, and then just as mysteriously dropped out of interest. See Arai, L. (2009). *Teenage Pregnancy: The Making and Unmaking of a Problem* (Bristol: Policy Press).

trajectory, to be alive and in good health until my oldest grandchild was five years old. When would I need to start childbearing? The answer for the poorest decile of English neighbourhoods: about 22. And that is almost exactly the age when people in those neighbourhoods do start childbearing, on average. As Arline Geronimus argued in a classic paper from years ago, young women who live lives of deprivation seem to know what they need to do, and so they do it.[17]

What about if I live in the median English neighbourhood? 28. Again, that's about what people actually do. And if I lived in the most affluent neighbourhoods? I could wait until after 30. And look; there we are. Enormous demand for IVF and egg-freezing, coming from rich neighbourhoods, because the rising health expectancies of the rich have prolonged the contextually appropriate schedule. That's fine; but let us not stigmatise the contextually appropriate behaviours of those who have to live their lives under different circumstances.

§

Isn't there a horrible circularity to this whole argument? You say that the poor smoke, don't adhere to medications, bear children young, and eat badly because they won't be alive long enough to see the negative consequences of these behaviours. But surely, the reason they won't live so long is exactly that they smoke, don't adhere to medications, eat badly, etc. So you seem in some way to be explaining their lifestyle by their lifestyle, which does not seem very satisfying.

This objection should not be hastily dismissed. When we do epidemiological studies of the relationship between social class and health or mortality, we always find that poor people fare worse than rich. Some of this is indeed because they are more likely to smoke. So you control statistically for smoking. Some of it seems to be due to poorer diet. So you control for diet. Some of it seems to be due to patterns of physical activity. So you control for physical activity. And it's true, the burden of excess mortality and morbidity is reduced by controlling for these things, maybe reduced by about a half. It is not reduced to nothing, though. However many voluntary behavioural things you control for, there is always a residuum of excess mortality

---

17 Geronimus, A. T. (1996). What teen mothers know. *Human Nature* 7: 323–52, https://doi.org/10.1007/bf02732898

and morbidity hanging over poor people. This, Gillian Pepper and I would argue, is the structural bit, the bit fundamentally due to too few material resources and too many demands, the bit that poor people cannot control except by not being poor (and if they had an available option of not being poor, we assume they would mostly take it up).

Our argument turns on this structural bit of health risk, this uncontrollable bit, being substantial. This structural excess health risk due to poverty is like predation and cold for wild mice; just there as part of the ecology, to be adapted to rather than opted out of. And you adapt to it by rebalancing between present and future consequences. Many of the arguments between left and right over the consequences of poverty are about the relative importance of the structural-ecological bit and the voluntary-behavioural bit of health risk. On the right, we decry people for being irresponsible, for not making better choices, not getting on their bikes to improve their lives. On the left, we are prone to point to structural sources of disadvantage, and invoke the criticism of having a victim mentality. The truth is that both bits are important. The account Gillian and I outline, though, hands an explanatory primacy to the structural-ecological bit. The presence of this structural increase in mortality and morbidity risk reduces the payoff for voluntary investments like adhering to medical recommendations, avoiding smoking, and so forth, and increases people's relative valuation of present enjoyment.

The voluntary-behavioural bit is important, though. In fact, it is responsible for a cruel irony I dubbed in an earlier paper the 'exacerbatory dynamic of poverty'.[18] Because of their structural-ecological disadvantage, the poor have less incentive than the rich to invest in their future health; but then the consequence of this reduced investment is to widen the health gap between the two groups to more than it structurally needs to be. And if the voluntary-behavioural choices of one generation partly determine the structural-ecological situation of their children, then we have scope for an inter-generational system of disadvantage that can self-perpetuate, and is hard to unravel. What is pretty clear, though, is that just putting larger warning labels on

---

18   Nettle, D. (2010). Why are there social gradients in preventative health behavior? A perspective from behavioral ecology. *PLoS ONE* 5: e13371, https://doi.org/10.1371/journal.pone.0013371

cigarettes and sweet foods is much like improving the sign-posting to the lifeboats on the Titanic. It is naïve for policy-makers or anyone else to assume otherwise.

§

Dickens' description of St. Antoine is characteristically evocative. Hunger and deprivation is written into the visual environment—in the ancient faces, the inadequate clothing, the dilapidation and litter, the poor foods on sale. All of these serve as 'grim illustrations of Want'. It is 240 years since the time Dickens was writing about, and 160 years since he wrote, but as I stroll around neighbourhood B, I muse on how little has changed. I instantly know that this is a poor neighbourhood, from the terrible litter, the state of the buildings, the clothes people wear, the things they are doing, as well as the headstones in the cemetery. The environment seeps information; information stares down from the chimneys, starts up from the kerb; it is written in the shops, the houses, the gardens.

In recent years I have become interested in the information that is freely available just by being somewhere. For me as a researcher, it is a resource. These days we under-do the simple acts of observation, the collection of this free information, which is why I am frequently to be found here counting passers-by, documenting whether doors are open or closed, tabulating litter or recording ages from headstones. These simple acts get neglected in an era of standardized surveys, controlled experiments, big data, focus groups, and discourse analysis. This is a shame—whatever your research predilections, it seems to me that the point of departure for research should always be the organism in its environment. Indeed, the organism in its environment (or individual in their context) is a shared starting point that unites behavioural biology and social science. So you can't do much better than put yourself in the environment, and ask: what do my study subjects see every day? What do they hear? What do they smell? Collect the information they collect, and it might help you begin to understand what they feel, why they do what they do.

It is not just the researcher who needs to harvest information. It is the study subject too. We come into this world with, within important limits, fairly open priors about what it will be like. So we have to detect our local environment, and cut our behavioural cloth accordingly. For this reason, characterizing the types of information available to the organism, the cues it can use to calibrate itself, is an important theoretical focus in behavioural biology.[19] I would like to see more explicit consideration of it in social science research too: the precise quantification of the information freely available to people in their daily lives, in their ordinary social environment. We know that poor people have different attitudes about the future than the rich, and Gillian and I have argued that these attitudes are contextually appropriate. But how do people know what is the correct attitude to develop for their particular ecology?

They are taught it, one might say. People tell them how they should behave. Or they imitate. Well, maybe, to a point. But I think there is a far greater role than we usually acknowledge for non-verbal inference based on sensory cues in the material environment. We know this is how visual perception works. What we receive is a set of cues of contrasts, surfaces and edges; what we infer is a world of objects and motion. By the same token, when we walk around neighbourhood B we see second-hand shops and litter and the mausoleum falling down (and, one morning as I ambled by, a wash-basin come crashing out of a closed upstairs window); what we receive is information. This is what life is like, will be like in the future. It doesn't need explicitly saying, or teaching, or pointing out. It is there, and you can no more not receive it than you can avoid perceiving a football as continuing to exist when it rolls behind a parked car.

This brings me back to the graves where I started. Every one of these deaths was a meaningful cue, never forgotten, to the living: to sons, daughters, siblings, friends. That's what could await me. Gillian carried out a study where she showed that more experience people had of bereavement, the more they devalued the distant future, and the sooner

---

19   See for example McNamara, J. M. et al. (2016). Detection vs. selection: Integration of genetic, epigenetic and environmental cues in fluctuating environments. *Ecology Letters* 19: 1267–76, https://doi.org/10.1111/ele.12663; Frankenhuis, W. E. and K. Panchanathan. (2011). Balancing sampling and specialization: An adaptationist model of incremental development. *Proceedings of The Royal Society B: Biological Sciences* 278: 3558–65, https://doi.org/10.1098/rspb.2011.0055

they wanted to start a family.[20] This makes perfect sense. The living can harvest information from death.

Perhaps this can answer another puzzle for us: the extraordinary elaboration of death in neighbourhood B. Although the cemetery itself is neglected and gracelessly dilapidated, many individual graves are tended and celebrated to a striking degree. Marble headstones that look remarkably expensive feature photographs, engravings of caravans, or pet dogs, or Newcastle United shirts. The flowers, balloons, bears, reindeer and cards are clearly renewed. And it is not just within the cemetery. It is very common as one walks around the West End to find flowers and cards tied to lamp-posts, railings or benches. Someone fell here. Someone loved this spot. I have not done a systematic study, but I don't believe you would find this degree of attention to death in a more affluent area.

There is only one conclusion you can come to: these deaths mean a lot round here. What does it mean for something to mean something? That's a rather involved philosophical question, but there are deep conceptual links between meaning, information, and uncertainty. A death means a lot if it carries a lot of information. And a death can only carry a lot of information if there is something about death we are uncertain about. For example, if every person died on the morning of their 79th birthday, there would be no information in age at death. We would be under no uncertainty about it. We would not say 'taken from us too soon' or 'sudden and unexpected loss'. We would not be shocked. But we have seen that in this cemetery, the standard deviation of age at death is 20 years. In other words, there is a lot of variation, probably more variation than would be true in an affluent place. Hence the neighbours live under uncertainty about when they are going to die. As a result, every death is informative. And when something is informative, you look at it for longer. You remember it for longer. You keep coming back here.

---

20 Pepper, G. V., and D. Nettle. (2013). Death and the time of your life: Experiences of close bereavement are associated with steeper financial future discounting and earlier reproduction. *Evolution and Human Behavior* 34: 433–9, https://doi.org/10.1016/j.evolhumbehav.2013.08.004

7. Why inequality is bad

> Macro-level data are characterized by inherent limitations in what they can tell us about individual-level processes.
>
> –Thomas V. Pollet and colleagues[1]

Richard Wilkinson and Kate Pickett's book *The Spirit Level: Why More Equal Societies Almost Always Do Better* caused something of a stir when it was published in 2009.[2] The thesis of the book is clear from its title. What many were struck by was the vast range of statistical evidence that the authors brought to bear in defence of their central claim. For outcome after outcome — life expectancy, physical health, mental health, crime, teenage births, social trust — they showed the same pattern. Economically equal countries such as Japan and those of Scandinavia have the best societal outcomes; the unequal USA fares badly; and there is a graded relationship across the countries in between. The average income of a country (as long as it is reasonably high) explains little of the variation in health and social problems; it is the inequality of the distribution of income amongst inhabitants that matters. Never can you have encountered a single explanatory factor that turns out to matter for so many outcomes, and, intriguingly, turns out to matter in exactly the same way for all of them.

The purpose of this essay is not to dispute that inequality is bad. I agree with Wilkinson and Pickett on this point — the evidence is incontrovertible, and they performed a major intellectual service in placing the issue of inequality so centrally on the political table. My interest lies more in their argument for *why* inequality is bad. They have a particular take on this, which we will get to below. It may have some merit. However, there is a more parsimonious alternative

---

1   Pollet, T. V. et al. (2014). What can cross-cultural correlations teach us about human nature? *Human Nature* 25: 410–29, p. 412, https://doi.org/10.1007/s12110-014-9206-3

2   Wilkinson, R. and K. Pickett. (2009). *The Spirit Level: Why More Equal Societies Almost Always Do Better* (London: Penguin).

explanation for their central results that merits equal consideration. What interests me is why they don't really discuss this alternative in their book (even to refute it), despite the fact that they must be well aware of it.

The centerpiece of Wilkinson and Pickett's evidence is a series of scatterplots, backed up with regression analyses, showing that more inequality (on the x-axis) goes with lower average levels of good stuff (like trust) or higher average levels of bad stuff (like mortality or crime) on the y-axis. Figure 2 reproduces a couple of them. The remarkable thing about these scatterplots is how similar they all look to one another; however diverse the outcome, you always end up with a roughly linear relationship, with some (aptly enough for a scatterplot) scatter, and a few interesting cases that look like they are doing a bit better or worse than you might predict. As a working scientist who understands how messy data are, I find myself crying out for one analysis that *didn't* work out that way. If it's a non-trivial association, it ought to sometimes *not* be there, or else you start to worry that it is somehow an artefact of the method. (Spoiler alert: it's not an artefact of the method, but it may be the inevitable product of a very general principle about money, as we shall see later).

The important thing to appreciate about these scatterplots is the following: the data points on them are not individual people. They are large aggregates of people, sometimes countries as in figure 2, but also in some of their analyses, US states. In some sense, this has to be so, because inequality is not a property of any individual person: it is necessarily a group-level property, exactly because it concerns how stuff is shared out across the social group. I have no objection to the idea that group-level properties such as the inequality of the distribution of national wealth affect the well-being of individuals. Clearly, they do. It's just that we have to be very careful about reasoning from statistical relationships that exist at the aggregate level, such as between countries' inequalities and their average health outcomes, and processes going on in individual bodies and minds.

Wilkinson and Pickett's explanation for the universal association between high inequality and poor welfare is an appealing one, and it is roughly the following. In societies where there are large gulfs between people, no-one can feel secure. Everybody is stressed: not just those at the bottom the heap, but also those in other social positions, who constantly need to feel worried about slipping down into penury, and feel they have to battle to hang on to their currently favourable position.

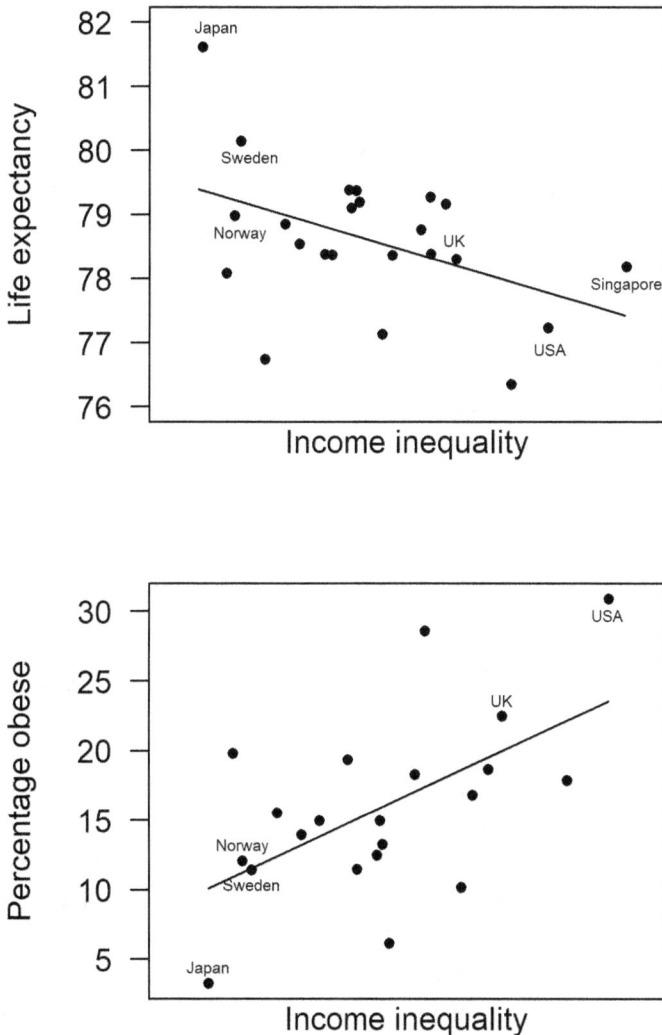

Figure 2. Scatterplots of (top) country-level average life expectancy against income inequality, and (bottom) percentage of adults who are obese against income inequality, redrawn from *The Spirit Level*. I have omitted most of the country names for clarity.

In short, everyone is worse off when there is more inequality because everyone is more stressed about either moving up, or staying where they are. In a more equal world, these kinds of stresses and motivations are relaxed, making way for more balanced and healthful approach to life. This in turn leads to fewer negative emotions, more trust and compassion, better mental health, better physical health, less crime, and so on.

Appealing as this narrative is, note what it has done. It has explained an association that exists at the aggregate level (comparing nations to each other) by a process in individual minds, by simply transposing the pattern we see when we compare groups into the head of every individual. As figure 2 shows, at the level of countries, you find poorer average well-being where inequality is higher; Wilkinson and Pickett explain this by saying that for every *individual*, their well-being goes down if the level of inequality in the surrounding society is higher. That could be true, but it need not be true to explain patterns like those in figure 2. We are entering the terrain here of the dreaded 'ecological fallacy' (the fallacy of assuming that an association at the aggregate level is reproduced within each individual), and related 'Yule-Simpson effect' (statistical relationships at one level of aggregation can be absent or even reversed at a different level of aggregation). Much has been written about these issues.[3]

The quantities on the y-axes of figure 2 and the other scatterplots in *The Spirit Level* are rates or averages for the country or state. Differences in rates or averages can come about in a number of different ways. To obtain the high rate of obesity in the USA compared to Sweden, for example, it could be that every individual within each country has a particular probability of becoming obese, and that probability is much higher for a person from the USA than a person from Sweden. This would be the Wilkinson and Pickett explanation: everyone in the unequal USA is at a higher personal risk of obesity, because of the stress of the surrounding inequality, than anyone in equal Sweden.

Here's an alternative explanation, though. Say there are two classes of people. The first class is people whose incomes are too low to buy good diets. Regardless of whether they live in the USA or Sweden, they have a 50% chance of becoming obese. The second class is people

---

3   See Pollet, T. V. et al. (2014). What can cross-cultural correlations teach us about human nature? *Human Nature* 25: 410–29, https://doi.org/10.1007/s12110-014-9206-3

whose incomes are high enough to afford good diets. Regardless of whether they live in the USA or Sweden, they have a 10% chance of becoming obese. These two classes are both present in both countries. Individuals of neither class are directly personally affected by the level of inequality in their country; all they need 'know' is whether they have enough money to buy a good diet or not, and this determines their risk of obesity. The difference between the countries arises from their different compositions in terms of the two classes. Sweden, say, consists of 95% people in the 'high enough income' category, and 5% of people in the 'not high enough income' category, whereas the USA consists of more like 50%: 50%. That could certainly produce the pattern shown in figure 2. And it's not a ridiculous explanation. There is lots of evidence that many people in the USA are too poor to buy decent food.[4] The thing about inequality is that it produces a big chunk of people who are really badly off given the general level of prices in their country. Smaller inequality produces a smaller chunk.

Wilkinson and Pickett could reasonably respond that they show, in still other scatterplots, that the average income of a country is not a very good predictor of health and social outcomes (given that we are comparing amongst countries that are all reasonably rich). Hence, it is not income per se that matters, but the inequality of its distribution. But again, those scatterplots are based on country-level *average* income not being a very good predictor of country-level averages or rates of health and social problems. What I am saying is that at the individual level, personal income might be very important—indeed causally the most important thing. This is quite compatible with a country's average income not telling you much about the average level of health, since the kind of income the matters at the individual level is not the national average income, but one's own.

§

As I mentioned earlier, the remarkable sensation one gets from reading *The Spirit Level* is how uniform a picture emerges from these scatterplots.

---

4   See: Gundersen, C., B. Kreider and J. Pepper. (2011). The economics of food insecurity in the United States. *Applied Economic Perspectives and Policy* 33: 281–303, https://doi.org/10.1093/aepp/ppr022; Ratcliffe, C., S. McKernan and S. Zhang. (2011). How much does the Supplemental Nutrition Assistance Program reduce food insecurity?. *American Journal of Agricultural Economics* 93: 1082–98, https://doi.org/10.1093/ajae/aar026

Every single one shows the predicted positive or negative correlation. It becomes an almost incantational moment, repeated throughout the liturgy: the unveiling of the scatterplot. You can't help but feel that these plots, diverse as they are in their data sources and outcome variables, must be revealing a principle of great generality. They are: the question is, exactly what principle is it?

The alternative to Wilkinson and Pickett's 'inequality around makes us all stressed' explanation is the following. At the individual level, income has diminishing returns for the outcomes that matter in life. When you put it like that, it is obvious. For a man who is starving, £10 can be the difference between living and dying; for a man who is rich, it is a bagatelle. For a man on a low income, £100 a month increase in income can be life-changing because of the material improvements he could make. For a university professor such as myself, £100 a month increase in income would, to be honest, not change my life in any very appreciable way.

What does this principle—the diminishing welfare returns to income—have to do with why inequality is bad? Here we need to think hard about what happens as societies become more unequal. The income inequality of a country can be thought of as a measure of the dispersion of the income of its individual inhabitants around the average. Where inequality is high, the dispersion is large. Where inequality is low, the dispersion is smaller and every individual is tightly clustered around the country average. Thus, the only way for a country to become more unequal whilst maintaining the same average income is for the dispersion to increase around a fixed central point: some individuals have to move away from the average income in the positive direction, while others have to produce an equal and opposite moment by moving away from the average income in the negative direction.

Now let us combine this principle with the idea that there are diminishing returns to having more income. Let us say that the relationship between an individual's income and his or her expected health (and here you can substitute, trust, stress, anxiety, probability of teenage conception, probability of committing a crime, any of the outcomes you like) is as shown in figure 3. This just puts onto a simple function the intuition that each increment of £100 in income is a little less beneficial for health improvement than the previous £100.

Now imagine we are going to increase the income inequality of the country without changing its average income. To achieve this, one group of individuals will have their income pushed away from the average in the positive direction, while another group will exactly offset this by having their income pushed away from the average in the negative direction. But, critically: the people who are pushed up in income will experience only a very small improvement in their health, because they are being pushed up across a zone where the income-health relationship is rather flat. By contrast, the people pushed down will experience a larger deterioration in their health, because they are being pushed across a zone where the income-health relationship starts to get steep. So the rich will get richer, but not much healthier, and the poor will get poorer, and much less healthy.

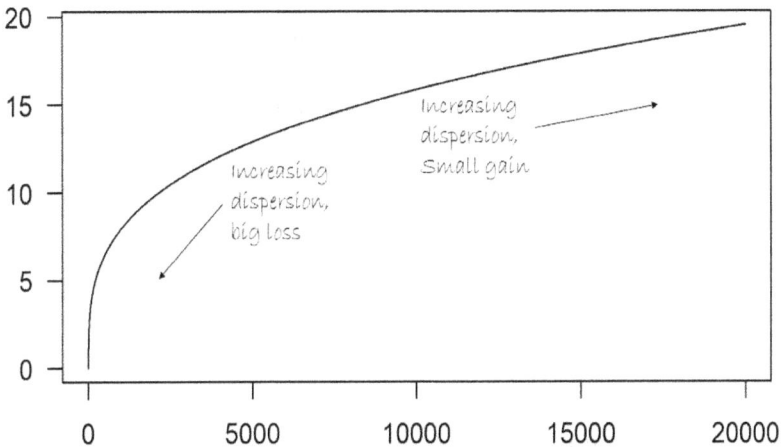

Figure 3. When there are diminishing health returns to income, increasing dispersion of incomes around a constant average produces a big health loss for the losers but only a small health gain for the gainers. Health is shown on an arbitrary scale.

So what will the average health be like for our hypothetical country after it has become more unequal without changing its average income? Well, its average health will be given by the following equation:

Average health after becoming more unequal =

The average health it had before +

a small improvement for those who did well in the increasing inequality −

a big deterioration for those who did badly in the increasing inequality

I hope you can see that 'the initial thing + a small thing – a big thing' has *got to be* of a lower value than the initial thing. In other words, if the relationship of income to health is of a diminishing-returns character, then it is necessarily the case that increasing inequality will make average health poorer. It is in fact a consequence of a general mathematical principle called Jensen's inequality.[5] For cases such as the obesity one, where the scatterplot shows a positive rather than negative trend, you have to assume that there are diminishing returns to income for avoiding obesity; then, the same explanation for the observed pattern then follows.

§

In case you are not yet convinced, I created a simple computer simulation. In my simulation, we study 30 countries, and from each we sample 100 inhabitants. The countries have the same average income per capita as one another (£10,000). They differ only in how unequally distributed this income is: the degree of dispersion around £10,000. Every individual's health (measured on an arbitrary scale) is determined by the square-root of their income (hence, diminishing returns), plus a sizeable dose of randomly-distributed noise.[6]

It is important to stress that in this simulation, no individual 'knows' the inequality of his or her country directly. They don't feel stressed by it, or fear falling into poverty, or feel the need to keep up with the Joneses. Their health is determined entirely by their personal income, plus random chance. Now, what happens when we compute the average health for each country, and plot it against the Gini coefficient, a standard measure of income inequality, for our virtual countries?

Figure 4 shows the results from four runs of the simulation. The association of higher inequality with poorer welfare is always there. The plots could have come straight out of the pages of *The Spirit Level*. Sometimes the association is stronger, sometimes a bit weaker, sometimes there are intriguing outliers or apparent non-linearities. But

---

5   Jensen's inequality: Roughly, the average of a function of X is not equal to that function of the average of X, unless the function is linear. See Denny, M. (2017). The fallacy of the average: On the ubiquity, utility and continuing novelty of Jensen's inequality. *Journal of Experimental Biology* 220: 139–46, https://doi.org/10.1242/jeb.140368

6   You can download the R code for running the simulation from: http://www.danielnettle.org.uk/inequality-r-code/

the point is: I have run this simulation hundreds of times. It is 17 lines of code (and I am not a very efficient programmer); there is no delicate psychology of shame and anxiety; no response of the individual to their psychosocial milieu; no representation of the society's Gini coefficient in the head of any individual; and yet we see *The Spirit Level*'s central result every time. That's mathematics for you. Under diminishing returns to income for some outcome, increasing dispersion in income will always decrease the aggregate level average of the outcome. This means any attempt to plot the correlation at the group level that is shown again and again in *The Spirit Level* is condemned to success by the laws of mathematics. No other assumptions about human psychology, stress, or anything else are required to explain this result.

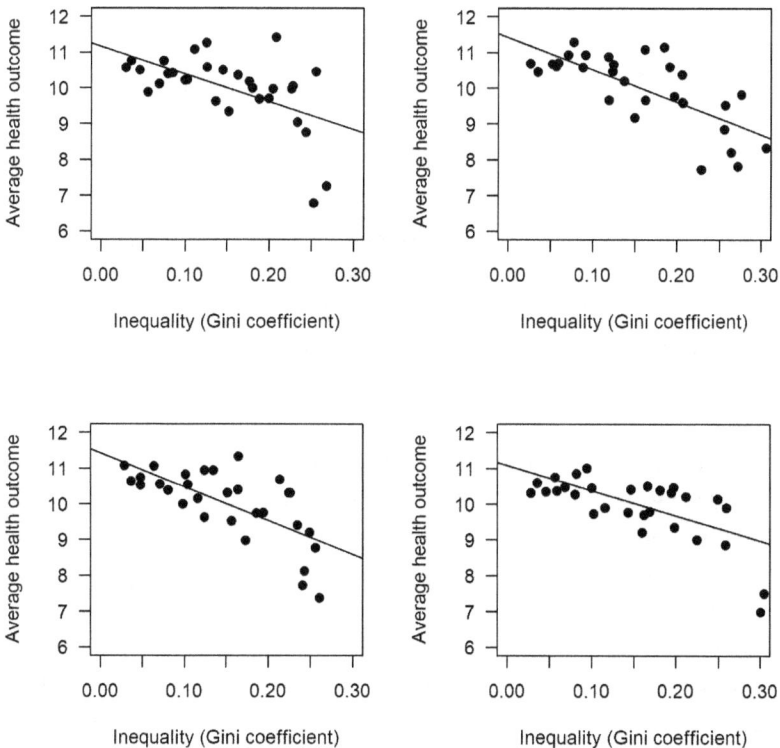

Figure 4. Income inequality against average heath for thirty simulated countries, four separate runs of the simulation, as described in the text.

We have two competing mechanisms that give rise to the same aggregate-level association: 1. More inequality in a society causes everyone in that society to feel worse; and 2. More inequality in a society produces winners and losers, and since the loss of the losers is bigger than the win of the winners, the average welfare becomes lower. Two competing mechanisms is good: this is the situation scientists like, because now you can try to test which is more important.

You can test between these mechanisms. One way is to gather what is called multilevel data. That is, you need datasets from multiple countries where the health and the income of every individual (not just the country average) is recorded. You also need to know the inequality of each country. Then you fit a statistical model that simultaneously estimates the effects of personal income and societal inequality on the individual's health. Second, and crucially, when you do this you need to specify that at the individual level, the relationship between income and health is likely to be non-linear, for example by relating health to the logarithm of income.

Researchers working on the association of inequality and health are well aware of the danger of the ecological fallacy, and for this reason, they have conducted studies using multilevel data. In general, they haven't done a very good job of estimating curvilinear relationships between income and health at the individual level. Epidemiologists seem keen on doing this by dividing income up into bands; this does allow for non-linearity, but dividing a continuum into bands always loses information, and therefore risks underestimating the magnitude effects of personal income. But anyhow, that's how they tend to do it.

In some studies, effects of societal-level inequality remain statistically significant after accounting for individual income, while in others they do not. For example, in two of three studies reviewed by Kawachi et al.,[7] there is no association between society-level inequality and individual health once the effects of personal income have been taken into account. In the third study,[8] there is some evidence for an effect of inequality on

7   See Kawachi, I. and B. P. Kennedy. (1999). Income inequality and health: pathways and mechanisms. *Health Services Research* 34: 215–27, p. 217. Download at: https://www.ncbi.nlm.nih.gov/pmc/articles/PMC1088996/

8   Kennedy, B. P. et al. (1998). Income distribution, socioeconomic status, and self-rated health in the United States: multilevel analysis. *British Medical Journal* 317: 917–21, https://doi.org/10.1136/bmj.317.7163.917

health after accounting for personal income, but, (a) the effects of societal inequality are much weaker than the effects of personal income; and (b) the negative effects of societal inequality on health after accounting for income are strongest in the poorer sectors of societies. In fact, there is no evidence for a negative effect of inequality on the health of the richest people in societies. Findings are similar in other studies. So when people model the problem properly, they have to conclude that, at the very least, a big part of the mechanism linking inequality to average well-being is simply the distribution of personal incomes.[9]

§

If I am leaning towards the 'diminishing returns from individual income' mechanism being the main one in accounting for the cross-national pattern, I fear you might think I am saying the following: 'Wilkinson and Pickett are wrong. Inequality is not really bad for societal outcomes. It is really just that there are diminishing returns to individual income. The group-level correlation is a kind of statistical artefact'.

On the contrary, I am saying: Wilkinson and Pickett are right that inequality is really bad for societal outcomes. Its negative effects may operate via a slightly different individual-level mechanism than they suggest. The group-level correlations between inequality and outcomes embody with the force of necessity a general truth that politicians should be aware of. Increasing inequality really is bad for societal outcomes exactly because there are diminishing returns to income. In general, in an affluent society, if you enact policies that make your society or group less equal, then well-being will get only a bit better for the winners, but well-being will get a lot worse for the losers. As a consequence, the

---

9 An even better approach than having multi-level observational data is using some kind of experiment or natural experiment. That is, we need to study a situation where people's experience of inequality is randomly varied without altering their actual incomes. It's hard, but there are some very interesting attempts to do it by studying the neighbours of lottery winners (Agarwal, S. et al. (2018). Does the relative income of peers cause financial distress? Evidence from lottery winners and neighboring bankruptcies. *Working Papers of the Federal Reserve Bank of Philadelphia* WP: 18–16, https://doi.org/10.21799/frbp.wp.2018.16), or by creating artificial micro-societies (Krupp, D. B. and T. R. Cook. (2018). Local competition amplifies the corrosive effects of inequality. *Psychological Science* 29: 824–33, https://doi.org/10.1177/0956797617748419). These kinds of research seem to me to give us the best window we can have on the causal properties of inequality itself.

aggregate well-being will be worse than it was before. Moreover, your social group will have to deal with all the knock-on costs of having sick, demoralized, hungry and desperate people around.

So, far from being a debunking of the idea that inequality is bad, this meditation leads me to affirm that inequality is bad. Indeed, it is potentially radical in its implications. The idea that the life returns to income might rather generally be diminishing is an argument in favour of reducing inequality (just as much as Wilkinson and Pickett's account is). More specifically, it is an argument for progressive taxation, and other redistributive mechanisms (taking from the rich, for whom it won't make much difference, and giving to the poor, for whom it could be life-changing). It is also a direct challenge to our prevailing cultural assumption that making a lot of money is a sensible and beneficial general aspiration for life.

I rather like the diminishing returns explanation for another reason. Accounts of why deprivation is bad for health have taken a rather 'soft' turn of late, in some quarters. If you want to read up on this, it's the debate between the psychosocial comparison account and the neo-materialist account of health inequalities.[10] The neo-materialist position is fairly obvious: disadvantage is bad for health because you can't afford decent housing, food, eye tests, and so on. For the psychosocial comparison camp, what's bad about disadvantage is that it produces feelings of shame, worthlessness and other negative emotions that arise when we compare ourselves to others.

It's true enough that experiencing disadvantage gives rise to negative emotions. You can document this in as many surveys as you like. As a causal decomposition of why poverty and deprivation might be harmful, though, the focus on negative feelings is rather bourgeois in its sensibilities. It almost says: I can't imagine that anyone is *really* poor in a country like contemporary Britain, not poor enough to actually starve or shiver or contract consumption or anything. It is just that, subjectively, coming last in the race makes people feel bad. But the idea that no-one in Britain has absolute problems of poverty — material rather than just

---

10   See Lynch, J. W. et al. (2000). Income inequality and mortality: importance to health of individual income, psychosocial environment, or material conditions. *British Medical Journal* 320: 1200–4, https://doi.org/10.1136/bmj.320.7243.1200 and a response by Marmot, M. G. and R. G. Wilkinson. (2001). Psychosocial and material pathways in the relation between income and health: A response to Lynch et al. *British Medical Journal* 322: 1233–6, https://doi.org/10.1136/bmj.322.7296.1233

subjective disadvantage—is a failure of imagination. If I may say so, it's a failure of imagination particularly easy to succumb to if you aren't poor yourself, and live in a neighbourhood where no-one is poor. In the West End of Newcastle, there are people, today, who do not have enough to eat. They don't have enough money to heat their homes or dress well. These are absolute material problems. Bad feelings are at most the canary in the coalmine.

So whilst both psychosocial experience and material wants may have important roles to play, I have a personal bias towards the neo-materialist approach. This is related to my gravitation towards the diminishing-returns-to-income account of the group-level correlation between inequality and societal outcomes. It is a little dismissive, even trivializing, of the experiences of poor people to characterize their difficulties as largely constituted by negative subjective feelings. Their problems—as they see them—are largely constituted by not having enough money to live on, given how much everything costs.[11] Similarly, I am uneasy with the argument that increasing inequality has hurt me, a university professor, as much as it has hurt a person on the minimum wage (I had to step over a homeless person on the way to the opera!). It hasn't really. The inconvenient truth is that many of the policies that have made the lives of poor people in Britain worse have been popular with the upper-middle classes, who perceive their lives to be made better, not worse, by them. The problem is that their gain has been more than offset by big losses for others. This raises questions of fairness, as well as the chaos of picking up the consequent social problems.

§

---

11   There is a wealth of research evidence supporting this claim. See for example: O'Brien, M. and P. Kyprianou. (2017). *Just Managing? What it Means for the Families of Austerity Britain* (Cambridge: Open Book Publishers), https://www. openbookpublishers.com/product/591, https://doi.org/10.11647/obp.0112; Daly, M and G. Kelly. (2015). *Families and Poverty: Every Day Life on a Low Income* (London: Policy Press), https://doi.org/10.1332/policypress/9781447318828.001.0001; Garthwaite, K. (2016). *Hunger Pains: Life inside Foodbank Britain* (London: Policy Press), https://doi.org/10.2307/j.ctt1t89f84; Taylor, A. and R. Loopstra. (2016). *Too Poor To Eat: Food Insecurity in the UK*, downloadable from https://foodfoundation. org.uk/wp-content/uploads/2016/07/FoodInsecurityBriefing-May-2016-FINAL. pdf.

Wilkinson and Pickett are very good researchers who are no doubt well aware of the theoretical issues—Jensen's inequality, the ecological fallacy, and so on—we have discussed so far. They don't talk about them in their book though. They just put forward their preferred mechanism and then give us scatterplot after scatterplot. This is despite the facts that: (a) diminishing welfare returns to individual income also predicts scatterplots showing the same pattern as their account; and (b) this has been widely known and agonized over in the technical literature for decades.[12]

I suspect this means Wilkinson and Pickett are smarter academics than I am, for two reasons. First: people in general (people like politicians, students, activists and the general public, as well as colleagues) will believe and reproduce an argument they find easy to understand and remember. Wilkinson and Pickett want you to believe and reproduce the idea that inequality is bad (as do I!). Wilkinson and Pickett's mechanism linking societal inequality to individual health is easy to understand and remember. The diminishing returns thing is not.

Humans find reasoning about certain kinds of processes hard. One kind of process they find particularly hard is that requiring what the great biologist Ernst Mayr called 'population thinking'.[13] Population thinking requires tracking the way groups of things (biological populations, societies) change their aggregate properties over time through changes in their composition, not changes to any of their constituent individuals. Population thinking is effortful to do. It is easier to conflate the properties of the individuals and of the group in one's mind, essentially to make the group itself a kind of representative individual.

For example, there is a large literature on popular understanding of biological evolution.[14] This literature shows people struggling to grasp

---

12   An early reference to relevance of diminishing returns to personal income for aggregate welfare is found in Melville, L. (1939). Economic welfare. *The Economic Journal* 49: 552–3, https://doi.org/10.2307/2224836, and the principle is a standard one in welfare and taxation economics. Its application to the present issue is discussed, inter alia, by Ecob, R. and G. Davey Smith. (1999). Income and health: What is the nature of the relationship? *Social Science and Medicine* 48: 693–705, https://doi.org/10.1016/s0277-9536(98)00385-2

13   Mayr, E. (1982). *The Growth of Biological Thought* (Cambridge: Cambridge University Press).

14   Shtulman, A. (2006). Qualitative differences between naive and scientific theories of evolution. *Cognitive Psychology* 52: 170–194, https://doi.org/10.1016/j.

that what happens in evolution is continual change in the proportions of individuals with different characteristics in the population, not necessarily change to any of the individuals themselves. The individual animals and plants are just being how they are; they don't know anything about the species or the direction of evolution. People find it easier to conflate what is happening at the individual level and at the species level. They believe that the species adapting to desert conditions over evolutionary time implies that individuals get better at living in deserts during their lifetimes (it needn't); or they endow the species itself with needs, drives, goals and so forth, properties that actually belong to individuals ('lemming X did something for the good of its species'). In short, people find it a lot easier to understand how one individual gets a longer neck by stretching it, than how the necks of giraffes could extend by a metre over evolutionary time without any individual changing their neck-length in their adult life, and despite offspring having the same length of neck as their parents on average.

The difficulty with population thinking crops up in trying to understand social phenomena, too. This may explain the widespread persistence of the ecological fallacy in social science research. I cannot tell you how many times I have tried to remember and transmit the following famous example. Amongst senators from the Northern states, a higher proportion of Democrats than Republicans voted for the 1964 Civil Rights Act. Amongst senators from the Southern states, a higher proportion of Democrats than Republicans voted for the 1964 Civil Rights Act. Amongst all the senators in the senate combined, a higher proportion of Republicans than Democrats voted for the 1964 Civil Rights Act. Eh? That can't be right. Actually, it is. It's quite straightforward. Most of the Republicans were from Northern states, where the support for the 1964 Civil Rights Act was high; more of the Democrats were from the South, where support was low. I can hang onto this one for about five minutes...no, it's gone again.

The difficulty with population thinking, I suspect, is why Wilkinson and Pickett's explanation for why inequality is bad has got such cultural legs. It says: the same thing is going on in the head of every individual as

cogpsych.2005.10.001; Nettle, D. (2010). Understanding of evolution may be improved by thinking about people. *Evolutionary Psychology* 8: 205–28, https://doi.org/10.1177/147470491000800206

is going on the level of whole groups. As the inequality gets greater, the individual=group feels worse. No population thinking required. What happens is the same for every individual in the group, and the group-level correlation is just an individual-level correlation. By contrast, the diminishing-returns requires non-linear functions, and averages being affected by some people moving in one direction with one effect, and other people moving another way with different effects. It's not catchy. It replaces a simple universal principle with a murky, nerdy statistical argument that could prove a cultural cul-de-sac. That does not help the cause.

The second reason that Wilkinson and Pickett's argument is a shrewd one is that it appeals directly to the self-interest of rich people. 'You too would be directly, personally better off if inequality were reduced', it says. 'Feeling stressed? Try lower inequality!' By contrast, the diminishing returns story says: 'you, the rich, might be slightly worse off under reduced inequality, but some other people would be much better off'. Wilkinson and Pickett offer us a free lunch for all; diminishing returns offers to make some (small) losers and some (big) gainers. Appealing to the self-interest of influential people is a smart move. It seems to have become much more normative in recent decades to assume that people can *only* be motivated to action by appealing to their self-interest. I doubt this is true: humans have complex moral emotions and perceived self-interest is certainly not the only factor influencing their endorsement of courses of action. But in the current climate, Wilkinson and Pickett's argument is pragmatically effective.

This is especially true since those who stand to gain most from reductions in inequality—poor people—have been systematically demonized, portrayed as undeserving, feckless, and *not like us*, in the cultural representations of recent decades.[15] This cultural situation risks making the argument that it is mostly poor people who would benefit from reducing inequality backfire rather badly. The argument that the rich would benefit too looks a better bet.

§

15   See Jones, O. (2016). *Chavs: The Demonization of the Working Class* (London: Verso).

I conclude this essay, as so often, in a state of profound unease. I find myself quibbling with the details of Wilkinson and Pickett's explanation of why inequality is bad, even though I admire their work and wish it to succeed in its broader social agenda. It would be easy, by raising complex alternative statistical explanations, to give succour to those who wish to dismiss the evidence that inequality is bad altogether. And there are plenty of those, such as Conservative former government minister David Davis, whose response was apparently: 'It's bullshit [...]. It's *bullshit* [...] I think it's one of those fashionable, stupid ideas. It's easy to sell a book, but I don't think it stands up'.[16]

If there is something you deeply believe, is it better to promulgate a simple explanation for it that people will immediately understand, support and remember, but is maybe a little simpler than what you, as a researcher, suspect to be really going on? Or is it better to be quite open about the uncertainties, the complexities, and the diversity of possible mechanisms that could underlie the phenomena we observe? Which one, ultimately, will increase the longevity and fecundity of the central idea? The media has for the most part made its mind up about this: I knew a TV producer who admitted to me that whenever an academic says, 'There are several possible reasons. First...', he quietly stopped his camera recording. No need to waste tape.

The issue is about the inherent complexity and uncertainty of scientific knowledge. As academics trying to make an impact on life, we could opt to express our doubt and uncertainty only behind closed doors, in lab meetings and the privacy of our studies; maybe in the technical literature, behind a screen of algebra. Then we could go out into to the world having decided what we believe, packaging it comprehensibly and appealingly, and defending it on all fronts with a greater certainty than we might privately hold. Or else, as I rather naïvely tend to do, we could stumble around *in* the world, incoherent and uncertain, with all our doubt, our qualifications, our evidential gaps, our multiple alternative mechanisms, our 'on the one hand's and changes of heart, in plain sight. It's a difficult problem.

---

16 Quoted in Jones, O. (2016). *Chavs: The Demonization of the Working Class* (London: Verso, p. 83).

8. Let them eat cake!

Of all the preposterous assumptions of
humanity over humanity,
nothing exceeds most of the criticisms
made on the habits
of the poor by the well-housed, well-
warmed, and well-fed.

–Herman Melville,
Poor man's pudding and rich man's crumbs

I had an interesting experience lately. Evidence shows that poor people tend to be somewhat more impulsive, anxious, irritable, and aggressive than rich people. I wrote a paper suggesting a hypothesis for why this might be the case.[1] Maybe, I suggested, they are just hungry. That is, at the time of completing the survey, or over the period of being observed, perhaps people with lower incomes are more likely to be hungry, or are hungry a greater proportion of the time, than richer people in the same sample. That could explain the observed correlations.

Hunger could explain the correlations because, quite separately, there are established literatures showing that when people—or animals of other species too—are hungry, they become more impulsive, anxious, irritable, and aggressive. In other words, hungry people show the very same suite of characteristics that is attributed to poor people in studies of socioeconomic differences. And the good thing about hunger is that we are not limited to correlation: we can manipulate hunger experimentally, within the same individuals, and show that hunger actually *causes* a shift to greater impulsivity etc. You just have to go without food for half a day. You can try it for yourself: it is one experiment I can pretty much guarantee will be successful. So, the hypothesis in my paper was based on two links: one from low income to hunger, and a second

---

1   Nettle, D. (2017). Does hunger contribute to socioeconomic gradients in behaviour?. *Frontiers in Psychology* 8: 358, https://doi.org/10.3389/fpsyg.2017.00358

    https://doi.org/10.11647/OBP.0155.08

from hunger to the suite of characteristics like impulsivity, anxiety, irritability, and aggression. The second of these links is absolutely rock solid, and unequivocally causal. The first link—that low income leads to greater hunger—relies on inference from more correlational types of data, but the evidence suggesting it is compelling too, as we will see. As I saw it then, I had taken a mystery—poorer people are more impulsive and irritable, for currently unknown reasons—and, by my hypothesis, demystified it into two steps, both of which we had prior grounds for believing to be correct. This left us with a simple, eminently testable scientific hypothesis, namely: the known relationships between income and hunger, and between hunger and impulsivity, irritability, etc., suffice to explain the observed correlations between poverty on the one hand and impulsivity, irritability etc. on the other.

So there I was feeling moderately pleased with myself. I didn't claim that my hypothesis was correct (we don't currently know that), but I did review the reasons for considering it, and discuss the ways it ought to be tested, without prejudice as to what the results of those tests would be. Feeling I had done what good scientists are supposed to do, I turned in my paper, and looked forward to the peer reviewers patting me on the head. Did they?

They did not. Generally, they hated it. More precisely, most of them hated most of it. I know because the scientific publishing industry gave me a number of opportunities to sample again from the pool of possible peer reviewers—as many samples as it took to finally find two who didn't hate it, or at any rate were generous enough to let the ideas get stated. Most of the reviewers who didn't completely hate all of it nonetheless wanted me to add multiple nuances of the 'there are likely to be many factors involved' kind. Authors of academic papers will be familiar with having to do this. You might think that such demand for nuance tends to improve theory, but Kieran Healy, in a robustly titled recent paper, has made a strong case to the contrary. The free-floating demand to add in more factors, he argues, 'typically obstructs the development of theory that is intellectually interesting, empirically generative, or practically successful', since it makes every theory more like every other one, and less easy to put to the test.[2]

2   Healy, K. (2017). Fuck nuance. *Sociological Theory* 35: 118–27, p. 118, https://doi.org/10.1177/0735275117709046

Of course, the problem may have been that I didn't write the paper well enough. But it felt like something more interesting was occurring. I have, often in my life, for various reasons, ended up writing bad 'on the one hand...' papers that blather on inconclusively about various esoteric topics without presenting any clear or socially important take-home ideas; papers that are all things to all readers; papers that contain nothing you could actually act on. The peer reviewers have usually loved these papers. To me, the hunger paper was worth much more than all those disappointing efforts combined. But the disappointing efforts are published in much more 'esteemed' journals, and seem to have been much better received, than the hunger paper. This is not an isolated case: Many academics will tell you that the papers they most value and are most proud of are the ones they have most trouble publishing, whilst their derivative, arcane or trivial ones sail through. I want to understand why.

§

Part of the reason for the poor reception of the hunger paper may be to do with intuition interference. I had an intuition that one particular experience—hunger—could be doing quite a lot of the work in explaining some of the subtle ways the poor behave differently from the rich. But I unrolled my idea on a terrain where others—specifically, the peer reviewers—already had quite developed intuitions about other constructs that could be important. 'What about parenting?', they asked. What about social norms? What about stress? What about the perception of relative disadvantage? Indeed. Some of these things might turn out to be nested within the hunger idea (there's a social norm in some social groups of being impulsive; but why did the social norm get established; perhaps because people in those groups are often hungry). Some are alternatives to my idea. But really, this reviewer reaction comes down to: 'my intuitions wouldn't have had me starting from there'. It's hard to know how to respond. You feel like saying: why don't you start from where you want to start from and see how far you can get, and I will start from where I want to start from and do likewise, then we will compare notes when we meet? But it's hardly a flaw of my paper that it is about what I want it to be about, not what you want it to be about.

A related refrain that arose in these discussions was: but surely a hypothesis that simple is likely to be wrong? Well, I agree. That's one of the reasons I felt so pleased with myself for having written the paper. Because science, I've heard, is about saying things that have the potential to be wrong. We even have a posh term for it: falsifiability. Essentially, science consists in making statements that have the potential to be judged, when all the evidence is in and the debates have been had, to be definitely wrong. If it couldn't be definitely wrong, it's pseudoscience. I have over-simplified here; I have articulated what is known to philosophers as naïve falsificationism. But more subtle positions in the philosophy of science still come down to scientific theories being different from non-science in their eventual 'discreditability' by the accumulation of evidence.

It seems to me to follow from this that the simpler your claim—the fewer constructs, linkages and relationships, the fewer degrees of freedom and reciprocal loops—the better you are probably doing in terms of scientific theorising. This point doesn't go down well in some of the waters in which I swim. People have a fondness for a kind of exhaustiveness in their theorising. They will earnestly present their 'theory' as a kind of flow diagram, with numerous boxes labelled things like 'parenting', 'social norms', 'perceived disadvantage', 'social comparison', and so forth. Pretty much every box has an arrow going to the outcome, and to pretty much every other intermediary box. Some pairs of boxes have reciprocal arrows. Some of the peer reviewers wanted me to change my paper from a statement of a simple hypothesis, to a review of the many factors likely to be involved in socioeconomic differences in behaviour, ideally with a ghastly diagram of the type described above as its figure 1.

Now the question is: in a thousand years, is my simple hunger hypothesis, or one of these many-factors-influence-the-outcome-and-also-each-other hypotheses, more likely to be left standing? The hunger hypothesis, presumably, is more likely to have failed. But rather than seeing that as a limitation, should we not see this as a good thing? There's a chance for the hunger hypothesis that in a thousand years we will be able to conclusively say: here's a possibility that people thought about, but turned *not* to be the answer (or not the whole answer). That's a kind of progress. For the many-factors theories, I think the most likely answer is that in a thousand years, as now, it won't really be possible to say whether they are still standing or not. This is because more or less any observation we make in the next thousand years is going to be compatible with such theories; weakly compatible, since the theories can accommodate so many slightly different patterns of covariance between the various things we measure. And many of the linkages in these theories are pretty much bound to be there (for example, poor people will always make negative social comparisons between their own situation and that of other people in society), regardless of whether there is any causal importance to them or not. I suppose I felt dismayed that the flaws laid at the door of the hunger hypothesis (its mono-factorial nature; its ignoring of many constructs currently discussed in the literature; its simplification of a complex reality) were exactly what I had most liked about it.

This is partly a question of taste. People vary in their tastes for stark simplicity versus swelling encampments or baroque twirls in explanations. And differences in taste, as Pierre Bourdieu observed years ago, often demarcate and reinforce fault lines between social groups.[3] A taste for simple models that can be exactly stated in a small number of equations demarcates many economists from their colleagues in most other social sciences. More generally, the veneration of theoretical simplicity versus reticulation constitutes a marked style difference between scientific communities. ("Hey Isaac. This gravity thing. There have got to be more factors than that in the motion of the planets... maybe motion itself *feeds back* to influence gravity through a nexus of reciprocal autopoiesis..."). As Bourdieu understood well, when you violate a distinction of taste, you can end up cast out from the social group that promotes that distinction. This is what kept happening to my paper.

§

As well as a general distaste for very simple hypotheses, my hunger paper seemed to tap into something more interesting; namely, an incredulity about the possibility that people in developed countries could really be hungry; that their hunger could be real and important, or could

---

3   Bourdieu, P. (1984). *Distinction: A Social Critique of the Judgement of Taste* (Cambridge, MA: Harvard University Press).

explain what they do. This incredulity appears to be widespread, and it doesn't stop at academics. Kayleigh Garthwaite nicely documents the incredulity amongst commentators and politicians in the contemporary UK, in her recent book *Hunger Pains: Life Inside Foodbank Britain*.[4] For example, faced with evidence of the massive increase in people in the UK relying on emergency food aid, former Conservative politician Edwina Currie was simply incredulous that people could actually be hungry. 'We should feel cross about [how many people are going to food banks], all of us' because '…they've just never learned to cook…', and, surreally, 'the moment they've got a bit of spare money, they're off getting another tattoo'. I know it's hard to understand, Edwina, but people with tattoos can be hungry too.

When I ask people to give reasons for their incredulity about people being hungry in affluent countries like Britain and the USA, they usually respond in one of two ways: 1. but poor people are often overweight, so they can't really be hungry; and 2. but they have *large televisions*. Both of these arguments are weak.

Argument 1: it is perfectly possible to be overweight all of the time, and hungry quite a lot of the time. Just think about it. Say that your cash-flow is very tight, so that for the later part of each week or month you don't have money to buy sufficient food and you go hungry, but when your benefits or wages arrive at the beginning of the month, you suddenly can buy food. What would you do? You would immediately go and buy as much affordable energy-dense food as you could, and you would quite understandably overeat. You've been hungry all week! You would probably buy food that is high in sugar and fat, because this gives you easily the most calories to the pound or dollar.[5] It could well be that, averaged over the whole month, you consumed more calories than you needed and stored some of your intake as fat; but nonetheless, for substantial parts of the month, you were hungry. It is well known that hunger and obesity tend to coexist within the same families, for exactly

---

4   Garthwaite, K. (2016). *Hunger Pains: Life Inside Foodbank Britain* (Bristol: Policy Press). Edwina Currie quotes from p. 68, https://doi.org/10.2307/j.ctt1t89f84

5   Drewnowski, A. and S. Specter. (2004). Poverty and obesity: the role of energy density and energy costs. *American Journal of Clinical Nutrition* 79: 6–16, https://doi.org/10.1093/ajcn/79.1.6; Jones, N. R. V. et al. (2014). The growing price gap between more and less healthy food: analysis of a novel longitudinal UK dataset. *PLoS ONE* 9: e109343, https://doi.org/10.1371/journal.pone.0109343

this reason; the extensive social-science literature on the topic refers to this as the 'hunger-obesity paradox'.[6] There is even some evidence that participation in food stamps programmes leads to greater weight gain, basically because the monthly timing of the arrival of the food allowance makes for a cycle of hunger and overeating.[7] The coexistence of hunger and fatness is not even a specifically human thing: if one group of birds is given constant access to food, whilst a matched group has its food taken away for periods of time, it is the group with constant access that remains thinner. The group with periodic hunger tucks in when it can, and stores extra calories as fat.[8]

Argument 2: Argument 2 is similar to argument 1 in failing to appreciate the temporal aspects of poverty. Just as being obese only means that *some of the time* you were able to buy enough calories to eat more than you expend, having a large television or other consumer good only means that *at least once in the past few years* you had a couple of hundred pounds to spare. And that's perfectly possible, since what characterises the precarious poor in affluent societies is not that they never have resources, but that their resources fluctuate close to the edge. All of us experience resource fluctuations; my bank balance is a couple of thousand pounds lower at the end of the month than the beginning. But in my case, the fluctuations are predictable, and anyway of no consequence to me, since I operate so far above the threshold where I would have to go hungry. A person experiencing less predictable fluctuations (for example due to inconstant employment or benefits

6   Dietz, W. H. (1995). Does hunger cause obesity?. *Pediatrics* 95: 766–7, downloadable from: http://pediatrics.aappublications.org/content/95/5/766; Scheier, L. M. (2005). What is the hunger-obesity paradox? *Journal of the American Dietetic Association* 105: 883–5, https://doi.org/10.1016/j.jada.2005.04.013; Nettle, D., C. P. Andrews and M. Bateson. (2017). Food insecurity as a driver of obesity in humans: The insurance hypothesis. *Behavioral and Brain Sciences* 40: e105, https://doi.org/10.1017/s0140525x16000947

7   DeBono, N. L., N. A. Ross and L. Berrang-Ford. (2012). Does the Food Stamp Program cause obesity? A realist review and a call for place-based research. *Health and Place* 18: 747–56, https://doi.org/10.1016/j.healthplace.2012.03.002

8   Ekman, J. B. and M. K. Hake. (1990). Monitoring starvation risk: Adjustments of body reserves in greenfinches (Carduelis chloris L.) during periods of unpredictable foraging success. *Behavioral Ecology* 1: 62–7, https://doi.org/10.1093/beheco/1.1.62; Witter, M., J. P. Swaddle and I. C. Cuthill. (1995). Periodic food availability and strategic regulation of body mass in the European Starling, Sturnus vulgaris. *Functional Ecology* 9: 568–74, https://doi.org/10.2307/2390146

delays) and/or operating closer to the edge, might well have the odd moment when things were looking better, and they then understandably wanted a TV to watch, but also lots of moments when fluctuations took them to the very edge. At these times they would have to go hungry. In fact, there is abundant evidence of poor people in Britain pawning their consumer goods during down-fluctuations in order to buy food, and having to buy them back again at inflated rates during up-fluctuations.[9]

So, in short, it is very easy to envisage patterns of resource fluctuations over time that would leave a person overweight and with a large television; and yet still often hungry, because their income was insufficient to buy food every day (see figure 5). But figure 5 is a hypothetical example. Is there empirical evidence that hunger is widespread amongst poorer people in affluent countries? This is where the rubber hits the road as far as my hunger hypothesis goes, of course; if there is not, then the hypothesis has no prospect of working. But the answer is that there is such evidence, and plenty of it.

In the USA, for example, social and nutritional surveys routinely measure the constructs of 'food insecurity' and 'food insufficiency'.[10] These are questionnaire measures based on items like 'In the last 12 months, did you or other adults in the household ever cut the size of your meals or skip meals because there wasn't enough money for food?' or, for households with children, 'Do your children ever say they are hungry because there is not enough food in the house?'. Based on these measures, around 16% of US households come out as food insecure, and 21% of children are classified either frequently hungry or at risk of being hungry. But, importantly, the percentages are much higher amongst those on low incomes: about 40% of households are food insecure, and 50% of children frequently hungry or at risk from hunger.

---

9   Garthwaite, K. (2016). *Hunger Pains: Life Inside Foodbank Britain* (Bristol: Policy Press), https://doi.org/10.2307/j.ctt1t89f84; O'Brien, M. and P. Kyprianou. (2017). *Just Managing? What it Means for the Families of Austerity Britain* (Cambridge: Open Book Publishers), https://www.openbookpublishers.com/product/591, https://doi.org/10.11647/obp.0112

10 See Nettle, D. (2017). Does hunger contribute to socioeconomic gradients in behaviour?. *Frontiers in Psychology* 8: 358, https://doi.org/10.3389/fpsyg.2017.00358, for more details of the evidence reviewed in this passage.

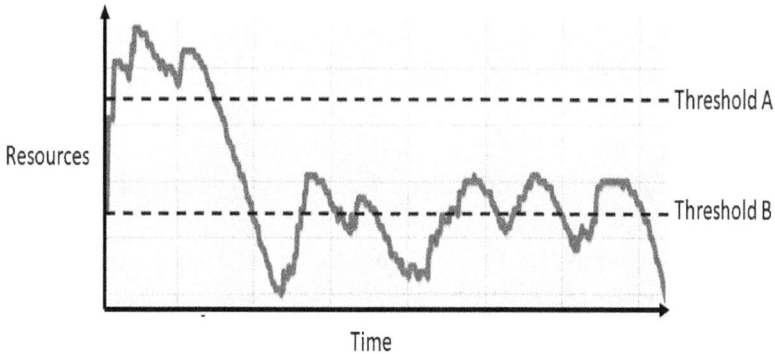

Figure 5. A hypothetical pattern of resources over time that would leave me with a large television and overweight, yet often hungry. Assume that when the money in my pocket is above threshold A, I can afford to go and buy myself a television or other reasonable consumer comforts. When it is between thresholds A and B, I can buy enough food to consume many more calories than I expend, especially by buying cheap energy-dense foods high in sugars and fats. When it is below threshold B, I can't afford to buy enough decent food to avoid hunger.

It doesn't stop there. Over forty-three million Americans are enrolled in the Supplemental Nutrition Assistance Program (SNAP); in other words, they receive food stamps because they cannot procure enough to eat. Yes, forty-three million: that's around 14% of the population. There is also a vast panoply of non-governmental food-assistance programmes and food pantries. In the UK, the largest single charitable food bank organization, The Trussell Trust, fulfilled 1.18 million referrals for three-day emergency food packages in the year to March 2017.[11] And although the Trussell Trust is the largest provider of emergency food aid, it is by no means the only one. 'Holiday hunger' is a widely reported problem amongst those on low-incomes. During school term, children receive free meals at school. In the holidays, the adults at home need to provide food for them. They may not be able to afford to do this, or in order to do so, they may need to go hungry themselves. In a recent survey of UK primary school teachers, 78% said they had seen

---

11 Information from Trussell Trust website: https://www.trusselltrust.org/news-and-blog/latest-stats/end-year-stats/

evidence that some children in their classes were going hungry during the holidays, and 37% said they had seen instances of malnutrition amongst children returning to school.[12] An all-party committee of the UK Parliament investigated hunger and food poverty in Britain in 2014 and concluded that hunger was a 'permanent fact of life' in the UK's poorest communities.[13]

I could go on, but I think you see my point. In the richest nations on earth, a lot of people are hungry a reasonable amount of the time. Those on low incomes are particularly likely to be hungry. Thus, if you sample a cross-section of the UK or US population at any moment in time, quite a few of them will be hungry, particularly those whose households are poorer. And that's all my hunger hypothesis needs.

§

When I started working on hunger, I assumed that the incredulity people have about the hunger of others was something specific to very affluent societies like twenty-first century Britain and the USA. In these societies we have so often been told, by the media for example, that our problems are problems of overabundance, that we just can't get our heads around the fact that this is not true for everyone. In earlier times or poorer countries, such incredulity would not exist.

As I go on though, I begin to appreciate that the incredulity might be a symptom of something much more general. After all, the original 'Let them eat cake!' was uttered by someone who lived at a time where famines were quite familiar, who had been told the peasants had no bread. Now, I say uttered by someone, because there is no evidence that it was really Marie Antoinette, the person to whom it is most often attributed. The phrase was actually put in the mouth of an unnamed 'grande princesse' by Jean-Jacques Rousseau in his *Confessions*. And what this great princess is supposed to have actually said is: 'Qu'ils mangent de la brioche!' which might be better rendered as 'let them

12   See   https://www.teachers.org.uk/news-events/conference-2017/ nut-survey-holiday-hunger

13   Quoted in Garthwaite, K. (2016). *Hunger Pains: Life Inside Foodbank Britain* (Bristol: Policy Press, p. 2), https://doi.org/10.2307/j.ctt1t89f84. See also: Taylor, A. and R. Loopstra. (2016). Too Poor To Eat: Food Insecurity in the UK, downloadable from https://foodfoundation.org.uk/wp-content/uploads/2016/07/ FoodInsecurityBriefing-May-2016-FINAL.pdf

eat brioche!'. But these textualities aside, the point is that even in a world where famine was a familiar occurrence, someone who was not hungry could not get her head around some other people in the same society being hungry. And this is not the only example. Apparently when Emperor Hui of China (over a thousand years ago) was told his people were starving because they had no rice, he reportedly said 'Why don't they eat ground meat?'. In the food riots that accompanied the Corn Laws of nineteenth-century England, magistrates were often keen to point out that the hungry rioters were not entirely destitute, and belonged to respectable trades, trying to suggest by this that what really drove them was not hunger, but avarice or malice.[14] This is Edwina Currie *avant la lettre*. Incredulity that others could really be hungry, or that hunger could be the real wellspring for their behaviour, seems to be general and long-standing.

This relates to a point made by George Loewenstein in a memorable article on 'visceral factors' in human decision-making.[15] By visceral factors, he means states like hunger and thirst, amongst others. Loewenstein's first, surely correct, argument is that these factors have a big influence on the decisions we make. His second argument is perhaps a more unusual one: when we are not in the grip of such states, we are not good at mentally simulating the decisions we would make if we were. We don't get it. A corollary is that when we ourselves are not in the grip of a visceral factor, we just can't understand the behaviour of other people who are. Why are they doing that, we ask? They've got tattoos!

I have become aware of the force of Loewenstein's argument in my everyday life. I periodically find myself near the finish line of a 10k road race or half-marathon, waiting to cheer my beloved wife home. Sometimes I can see that, in the final few hundred metres, she is not far behind a rival, or is in contention for a personal best time. When it's not clear whether she will prevail, I sometimes find myself thinking, 'Why doesn't she just…run a bit faster?'. I know she is capable of running a

---

14 Sutton, J. (2016). *Food Worth Fighting For: From Food Riots to Food Banks* (London: Prospect Books, p. 23).

15 Loewenstein, G. (1996). Out of control: Visceral influences on behavior. *Organizational Behavior and Human Decision Processes* 65: 272–92, https://doi. org/10.1006/obhd.1996.0028

bit faster. I almost resent it as a caprice that she doesn't, and I have to censor myself. But of course, in that moment, *I'm not fatigued*. The reason she doesn't run a bit faster is that she is exhausted. When I am running, I get exhausted too. But somehow, when I am not fatigued myself, the right intuition doesn't come to me, and all the wrong ones (Is she really trying?) come to mind.

This visceral-state-blindness relates to a classic psychological phenomenon known as the fundamental attribution error.[16] The fundamental attribution error refers to our default position, when confronted with people's behaviour, of attributing it to their personality or enduring dispositions, rather than their current situation. We systematically neglect the temporal fluctuations in people's states, in favour of assuming they are just always like that. The fundamental attribution error is poorly named, because sometimes, often even, it produces the correct attribution rather than an erroneous one. So it's better thought of as an explanatory style; one that may be reasonable or prudent on average, but in particular instances leads us to neglect powerful situational influences—such as hunger, itself a product of the powerful influence of having no money right now. That's why commentators are so prone to argue that reliance of food banks must reflect poor moral character, or poor planning, or that 'they never learned to cook', when I am afraid the true situational culprits are staring us in the face.

I don't know why we would be so bad at mentally simulating the influence of visceral factors on ourselves and others. It would seem to me very useful to be able to detect and anticipate such regularities in behaviour, but Loewenstein's argument suggests that we aren't good at it. And perhaps that's why the reviewers had such a problem with my hunger paper; an intuition blank around the possibility that a simple visceral factor could really be what is at work. Because really my paper was radical in its implications. You know poor people, it said; they're *just the same as you*, only hungry. If you were hungry, you would behave like them; and, most importantly, if we could ensure that everyone in society had secure daily access to abundant and nutritious food, these social differences would simply and instantaneously disappear. No

---

16   Jones, E. E. (1979). The rocky road from acts to dispositions. *American Psychologist* 34: 107–17, https://doi.org/10.1037//0003-066x.34.2.107

complex arguments about the culture of poverty; no arcane theories about anomie, epistemes or structuration; no feedback loops. The issue would just *go away*. As I said earlier, this hypothesis may well not be correct. It probably isn't correct. But it's an interesting and audacious claim that you could actually do something with. Why would it be bad to try to test it to destruction?

§

This has got me to thinking: are hunger and food systematically neglected, as topics of investigation or sources of explanation, across the contemporary human sciences? I suspect perhaps they may be. For example, celebrated cross-cultural studies have shown that subjects from different societies across the world vary in their behaviour in artificial social dilemmas known as economic games.[17] In these games, people can either contribute to a social good, or behave more selfishly, and others can choose (or not) to punish them for selfishness. In the cross-cultural studies, the observed behaviour varies substantially within the sample from any one study site, as you would expect, but there are also statistical differences between samples from study sites located in different societies. Of all the various ideas that have been put forward to explain the variation across study sites, I have not encountered the proposal that people in some study sites are on average hungrier than those in others at the time of testing. The idea makes some sense: the diverse sites studied subsist on everything from hunting and gathering to the Western diet, so there would surely be variation between as well as within sites in what, how much and how recently people have been eating.

Recently, my student Sam Fraser created laboratory micro-societies of volunteers, in which the participants were randomly assigned to either have breakfast as usual, or to skip it. They played one of the same economic games as used in the cross-cultural studies. Sam found that in the 'hungry' micro-societies as compared to the breakfasted ones,

---

17   Marlowe, F. W. et al. (2008). More 'altruistic' punishment in larger societies. *Proceedings of the Royal Society of London B: Biological Sciences* 275: 587–90, https://doi.org/10.1098/rspb.2007.1517; Henrich, J. et al. (2010). Markets, religion, community size, and the evolution of fairness and punishment. *Science* 327: 1480–4, https://doi.org/10.1126/science.1182238

there was less punishment of selfish behaviour, and as a consequence of this, more selfish behaviour went on.[18] The cross-cultural finding is that participants from smaller-scale societies are less likely to punish selfish behaviour; and where there is less punishment of selfish behaviour, more selfish behaviour goes on. All that would be needed to use Sam's (admittedly preliminary) finding to explain the cross-cultural ones is a demonstration that participants from smaller-scale societies are more likely to be hungry at any given moment, which does not seem an entirely unreasonable idea, given that such societies are generally impoverished and disconnected from the global food system. But as far as I am aware, the many variables measured in cross-cultural studies do not include what the participants had for breakfast. Researchers leap instead for explanations that are less visceral, but also less grounded in what we actually know of how the individual, embodied human decision-maker functions.

This is just one example of food-blindness in the contemporary human sciences. If you start going on about the hunger drive and control of feeding, as I am prone to do, you feel like a behaviourist throwback to the 1950s. The textbook models of human cognition hardly seem to consider hunger as an input or eating as an output. Hunger and eating are seen, perhaps, as marginal or low-level 'biological' processes, barely even cognitive, not as interesting as spelling words or working out how likely people are to become bank tellers. In the vast literatures on human judgement and decision-making, the judgements and decisions studied are often about money, but almost never about food. This is curious, because most human societies through history have lacked money, but not a single one has lacked eating. A good argument can be made that the mechanisms with which we make monetary decisions actually evolved to deal with food options. It's only by a secondary exaptation of them, in contemporary societies, that we can decide between financial dilemmas. It's not clear to me why food and eating aren't more central topics. Many a times a day, you make life-and-death decisions about what to put in your stomach, and generally you do it so remarkably well that the process goes unnoticed. When was the last time your survival

---

18   Fraser, S. (2018). *Effects of hunger on human cooperation*. (MRes dissertation, Newcastle University).

depended on a correct inference about whether someone was a bank teller or not?

Faced with all this, it's hard to avoid the conclusion that the contemporary human sciences are written by the well-fed. Hardly experiencing the state of hunger, academics can't imagine hunger and food as central issues in human life. Hungry? You just go to the canteen, neat and quick, then you can get straight back to work on social identity. A particularly disappointing non-player at the table here is the paradigm known as evolutionary psychology. With its keenness on relating contemporary psychological processes to their evolutionary origins, and exploring continuity with other species, you would think evolutionary psychology would make hay in the fields of hunger and eating. But there is really rather little work done on the topic.

This is rather odd, given that evolutionary psychology can hardly be accused of neglecting another visceral factor: sex. It sometimes feels like evolutionary psychology is mainly about sex, in particular the choice of how and with whom to have it. I don't know much about your ancestors, but what I can be sure of is the following. Each of them managed to have sex, with someone of approximately the right species, at least once in their lives. By contrast, they had to procure and select several thousand calories of appropriate food *every single day*, never starving and never poisoning themselves, for tens of years. Put like that, what ratio of evolutionary psychology research papers on the psychology of hunger and food to the psychology of mate choice and sex, ought we to expect? And look at the ratio we observe.

What this tells us is that evolutionary psychology, so far, is mostly the cultural invention of affluent college students and those who interact with them. Perhaps when you are nineteen, privileged and live on a college campus, you are rather more concerned about who your next sex partner will be than where your next meal is coming from. You can't imagine this to be anything other than the normal state of affairs for humans. And so there is a great deal of evolutionary psychology research about dating and hooking up, and then a bit about stuff like making friendship groups and working in teams; and not a lot about hunger, poverty, domination, social conflict, infirmity, death. This may be part of the reason for evolutionary psychology's image problem amongst social scientists. Social scientists' (philosophically unnecessary) dislike of evolutionary psychology is partly founded on (sometimes wilful) mischaracterisation and misunderstanding of its premises, as has been well discussed.[19] But another part of it is simply due to evolutionary psychology's topical obsessions, which seem frivolous to those who work amongst poor, ageing, threatened or socially marginalized people, or on pressing societal issues. I have been trying to argue for a number of years that the best thing evolutionary psychology could do for its image problem would be to show up at the debates about poverty and inequality within our affluent societies.[20] And if that means we don't get so much time to worry about optimal breast size or the significance of how far apart one's eyes are, I for one would accept that as collateral damage.[21]

Alright. Here endeth the lesson. Enough self-righteousness from me. I need lunch.

---

19   For example by Kurzban, R. and M. G. Haselton. (2010). Making hay out of straw: Real and imagined controversies in evolutionary psychology. In J. Barkow (ed.), *Missing the Revolution: Darwinism for Social Scientists* (Oxford: Oxford University Press, p. 149–66), https://doi.org/10.1093/acprof: oso/9780195130027.003.0005

20   Not all evolutionary psychology deserves the criticism I have voiced here. Special mention to venerated pioneers Martin Daly and the late Margo Wilson, e.g. Wilson, M. and M. Daly. (1997). Life expectancy, economic inequality, homicide, and reproductive timing in Chicago neighbourhoods. *British Medical Journal* 314: 1271, https://doi.org/10.1136/bmj.314.7089.1271

21   Though for the record, your assessment of optimal breast size seems to depend on how hungry you are: Swami, V and M. J. Tovée. (2006). Does hunger influence judgments of female physical attractiveness? *British Journal of Psychology* 97: 353–63, https://doi.org/10.1348/000712605x80713

9. The worst thing about poverty is not having enough money

> If the poor are poor due to bad choices or preferences, then providing them with additional income alone will not necessarily achieve any observable improvements
>
> –Randall Akee and colleagues[1]

In his 2015 speech to the Conservative party conference, then-Prime Minister David Cameron vowed to use his remaining time in office to mount an all-out assault on poverty in the UK. A worthwhile aspiration, indeed; and not an aspiration we necessarily expect to hear from Mr. Cameron's side of the political spectrum. As it turned out, Mr. Cameron's remaining time in office was not to be very long. In less than a year, he had burnt his wings in the EU referendum and disappeared without trace. I want to talk about an interesting feature of his anti-poverty evangelism, though: central to his planned assault was the idea that poverty was not entirely, perhaps not even mainly, about money.

The intellectual work behind the Cameron approach to poverty was carried out in the preceding years, primarily by the Centre for Social Justice (CSJ) think-tank.[2] The CSJ's analysis is, like the curate's egg, good in parts. The CSJ quite rightly stresses that low incomes are correlated with a whole raft of non-income problems. Low-income families are disproportionately likely to be affected by: addiction; alcoholism; family instability; criminality, anti-social behaviour; educational failure; and

---

1   Akee, R. K.Q. et al. (2015). How does household income affect child personality traits and behaviors? *NBER Working Paper* No. 21562, p. 16, https://doi.org/10.3386/w21562

2   See for example their 2012 policy paper 'Rethinking child poverty' (https://www.centreforsocialjustice.org.uk/library/rethinking-child-poverty) and their 27th March 2013 blog entry 'It's not all money, money, money' (https://www.centreforsocialjustice.org.uk/csj-blog/its-not-all-money-money-money).

    https://doi.org/10.11647/OBP.0155.09

so on. So there is a manifold of social issues that cluster together, and make life unpleasant or difficult for certain parts of the population. The CSJ rightly argues that if you *just* raised poor people's incomes, whilst making no impact at all on the unequal burden of these other problems, you would not have cracked the problem of social disadvantage in its entirety.

The CSJ then proposes that we *measure* poverty, not just by the amount of money people have, but by a basket of indicators including all these other things like alcoholism, family instability, and so forth. This proposal has, as far as I can see, no merit whatever. It is one thing to acknowledge that poverty is correlated with all kinds of non-income issues. Maybe it is even causally connected to those other issues. But the best way to guarantee that you will never be able to tease out the linkages is to measure poverty in such a way that confuses it with the other issues at the outset. Let me take an example: suppose I am interested in how ocean temperature relates to coral bleaching. Because I feel these things are linked, I could propose to measure ocean temperature by a raft of different indicators including the extent of coral bleaching. The one thing I would now be unable to do is find out whether ocean temperature is related to coral bleaching. I have simply muddled them by assumption; having done so, it becomes impossible to study the relationship between them, because you can't even identify the two phenomena you wish to relate. Thus, whilst I and many social scientists would concur that *well-being* is not just about income, claiming that *poverty* is not just about money is a bit like saying that hyperbolas are not just about a plane intersecting both halves of a double cone. Isn't that, kind of, how you know you are talking about a hyperbola rather than something else?

If we set aside the CSJ's definitional peculiarities, though, we see that there is an interesting idea in there somewhere. Poverty, they say (presumably with the income definition of poverty in mind in this instance), is often 'a symptom of deeper social issues'. What do we mean when we say 'a symptom'? Typically, a symptom is: (a) one of a network of associated phenomena, as in 'symptoms include swelling, fever and rash'; and (b) by implication, not the one you want to go for if you want to causally manipulate the system, as in 'it's best to treat the cause rather than just the symptoms'. So really, the CSJ is making an empirical claim, namely: if you want to lessen the well-being burden due to the inter-related network of poverty, family breakdown, addiction, and so forth, then raising income is not the most effective strategy. Instead, we need to tackle the other nodes directly. Incomes will follow in turn, as better-functioning families get their lives into order and become more economically productive. In fairness to the CSJ, this is hardly a laissez-faire recipe for benign neglect of poor people. It gets the government off the hook in terms of the moral case for direct redistribution of cash. But for the government seriously to take on the mantle of responsibility for the family relationships, narcotic consumption, educational attitudes, and normative behaviour of every individual in the land is a mind-blowingly interventionist, not to mention very expensive, aspiration to hold.

The CSJ then, has put out there a big idea. No problem with that. It's just that there is a growing consensus in social science for the opposite view: if you want to deal with the manifold of social problems faced by poor people, both here in the UK and in developing countries, just giving people money is actually a pretty effective strategy. Accepting this opposite view does not come easily to me. I attained my political consciousness in a third world development movement which was pretty much predicated on the aphorism, 'give a man a fish, and he will feed himself for a day; teach a man how to fish, and he will feed himself for a lifetime'. It's hard for me to accept that just giving out fish can possibly be right. I am going to spend the rest of this essay reluctantly conceding that it could be.

§

Let's all admit, for the sake of argument, that low income, family breakdown and addiction are related to one another. I don't just mean that they are correlated. I mean that there are real causal linkages from each of them to both of the other two. Low income increases the likelihood of developing addiction, and of families breaking down; addiction increases the likelihood of family breakdown, and of losing income; and family breakdown increases the likelihood of losing income and of developing an addiction. It's a mutually reinforcing trio of problems: a dynamical system. Now let's say you want to make the world a better place. Where would you do best to put your dollar? You can choose between directly raising incomes; providing addiction treatment programmes; and providing family counselling.

One of the things you need to consider is the magnitude of the effect of changing one of the variables on each of the other two. For example, if you can reduce family breakdown, to what extent do income and addiction then improve? The CSJ hypothesis is, in effect, that the knock-on impact of reducing family breakdown or addiction for income would be rather large, but the effect of raising income on the other two problems would be small. Perhaps it would even be zero, or negative, as previously poor people went out and frittered away their newly-acquired cash on social bads like drugs. *They wouldn't know how to use the money sensibly.* So naturally, the CSJ concludes that raising incomes alone is not the best approach.

The examples we are going to see suggests that they have it the wrong way around. Raising the incomes of poor people, even absent any other changes, can have a surprisingly large positive impact on all kinds of social and behavioural problems, and hence well-being. It does not eliminate all social problems, of course: nothing we know of does that. Nonetheless, it can do a lot to reduce the non-income wellbeing disparity between rich people and poor people, as well as, more obviously, the income disparity. It makes sense that, other things being equal, raising incomes is likely to be the most effective way of perturbing the dynamical system of social and family problems. That's because giving people cash is remarkably efficient, especially if you do it in some fairly non-bureaucratic way. There's a few cents in the dollar for administration and banking charges, but beyond that, the more money you transfer to poor people, the more their incomes go up. The efficiencies of family counselling and drug treatment programmes are likely to be much lower. I am not saying these initiatives don't work at all; I am sure they do. But you have to recruit and train up counsellors and staff. These people are typically much more middle-class than the people we are trying to target. They need decent compensation packages, and that costs a lot, typically much more than a poor family earns. For overseas development, they need to be flown in and housed. Then they have got to access the populations with the need. And even assuming they manage to do this, their help only has a certain degree of success; plenty of families go through family counselling and still break up anyway; plenty of addicts receive treatment but don't escape their addiction. So it would probably be fair to speculate that the efficiencies of non-income forms of aid directed at poor people are typically lower than that of direct income support.

§

The same people and places tend to have the lowest incomes, the poorest physical and mental health, the most crime, the lowest trust, more behaviour problems, and so on. However, this does not in itself help you decide on the best remedy for poverty. Both the CSJ and the cash-first hypotheses are consistent with there being a manifold of positive correlations of all the different kinds of life-crapness. If you want to get anywhere in adjudicating between the two hypotheses, you need something like the scientific experiment. In an experiment, you hold everything else constant, and perturb one variable (for example, income) in the absence of any other change. Then you see what effects follow on the outcomes that interest you. Hold on, you say, that's all very well. But social scientists can't do experiments. People's incomes never change without their education, culture, or other aspects of their behaviour changing too, in uncontrolled ways. Social, political and economic life just go on, and we social researchers are limited to documenting them and interpreting their fluxes.

The situation is actually not quite as bad as this. Sometimes one factor does get changed, pretty much independently of all the others, and for reasons that are largely exogenous to the system. Social scientists spend a great deal of time studying these situations, and the results come as close to a decomposition of causality as you could reasonably hope for. The gold standard scenario is the *randomised control trial*, the true scientific experiment applied to a social policy innovation. More and more of these are now done. But even where randomised control trials have not yet proven possible, there are nearly-as-good sources of causal inference: *natural experiments* or *quasi-experiments*. These are situations where some change occurs that is outside the researcher's control (this is how it differs from a true experiment), but nonetheless alters just the variable of interest, and just for some people but not for some other, comparable ones. When a social policy is introduced into one jurisdiction but not a similar neighbouring one, then as long as the reason for the introduction happening where it did is not reducible to any existing characteristic of the jurisdictions, then you have a kind of natural experiment. And social policy changes sometimes happen for the strangest and most random of reasons.

My favourite quasi-experiment comes from the Great Smoky Mountains study. This began as a fairly run-of-the-mill longitudinal study of psychiatric problems, addictions and problem behaviours amongst young people in parts of Western North Carolina, beginning in 1993, and continuing as the young people grew into adults. But it became something far from run-of-the-mill in 1996. A fair proportion of the participants were Native Americans from the Eastern band of Cherokee. In 1996, a casino was opened on their reservation land (Native American reservations are outside state gaming laws). Some of the profits were put back into the band community, and the mechanism chosen for doing this was basically a Universal Basic Income: all adult band members received an equal portion, in the form of semi-annual cash payments, for which they did not have to do anything other than be themselves. Small at first, these payments had risen to $9000 per person per year by 2006, enough to very substantially raise household incomes in that part of the world. And for Eastern Cherokee youth, there was a large lump sum to be held in trust and collected on their 18th birthdays.

It's important to appreciate that, before the payments began, the Eastern Cherokee had the usual poverty smorgasbord: as well as their incomes generally being low, there were lots of problems of addiction, anti-social behaviour, and family strife. It was classic CSJ stuff. And if the CSJ hypothesis were right, then the cash payments, which after all did nothing at all but lodge a cheque, would not have helped with all these other 'symptoms of something deeper'. Things could have even got worse. Suddenly having cash in the bank, and lacking the family stability and life skills to know what to do with it, you might have expected the newly cashed-up young people to drop out of school (who needs to work when you are given money for nothing?), and turn to drink, drugs and gambling. Nothing could be further from the truth.

There are several good studies of what happened to the Eastern band of the Cherokee, so here I will focus a few of the most noteworthy. Elizabeth Jane Costello and colleagues systematically compared young men and women from Cherokee families with non-Cherokee of the same age from the Great Smoky Mountains cohort.[3] These non-Cherokee were effectively the control group. Not a very good control, you might say, since the non-Cherokee were bound to differ from the Cherokee in many non-income ways. However, the researchers could turn here to the fact that they had data from Cherokee of different age cohorts. The oldest cohort had benefited rather little from the casino scheme—the lump sum payable at 18 only started to cumulate in 1996, so those turning 18 in 1998 got only a very modest amount, and had not benefited from increased parental income for very long either. So the differences between the oldest cohort of Cherokee and oldest cohort of non-Cherokee tells you something about the *status quo ante casino*. By contrast, the youngest Cherokee, turning 18 in 2002, received $35,000 on their birthday, besides which their parents had had quite large sums coming in for all of their teenage years. So if cash does anything good for non-income outcomes, you should see the youngest cohort of Cherokee doing better relative to their non-Cherokee peers than earlier cohorts of Cherokee had done. This is a variant of what is called a 'difference in differences' study design, because any causal impact of the money is going to change the differences between Cherokee and non-Cherokee outcomes between the oldest cohort (not much casino cash), and the youngest cohort (lots of casino cash).

And the differences were indeed different. Looking at the oldest cohort, by the time of study, 41% of Cherokee had experienced some kind of psychiatric disorder, against 31% of non-Cherokee. Much of this was made up of or included some kind of substance dependence (35% of Cherokee, against 29% of non-Cherokee). The rates of diagnosed 'behavioural disorder' (which is often a catchall for minor criminality and anti-social behaviour) were five times higher in the Cherokee than the non-Cherokee. But remember these were the Cherokee cohort who had benefited only marginally from the coming of the casino. In the youngest cohort, who had benefited very substantially from casino money, not only had the Cherokee caught up with their non-Cherokee brethren, but they had surpassed them. The differences were all in the opposite direction: any psychiatric disorder: 31% Cherokee versus 37%

3   Costello, E. J. et al. (2010). Association of family income supplements in adolescence with development of psychiatric and substance use disorders in adulthood among an American Indian population. *Journal of the American Medical Association* 303: 1954–9, https://doi.org/10.1001/jama.2010.621

non-Cherokee; substance dependence: 23% Cherokee against 35% non-Cherokee; behavioural disorders three times higher in the *non*-Cherokee than the Cherokee.

In related work, Randall Akee and colleagues looked at involvement in criminal activity, and at school performance, whilst the members of the study were still minors.[4] Again by comparing those who received different amounts of casino transfer and those who received none, they were able to estimate that an additional $4000 per year of unearned income per year reduced the likelihood of ever getting involved in minor crime by 22% (for a 16 or 17 year old); and, moreover, that it increased the average amount of formal schooling completed by a whole year. Prior to the casino, Cherokee youth had worse rates of minor criminality and lower rates of high school completion than non-Cherokee youth. Over the first few years of the casino, they not only closed the gap, but went beyond: now they were *more* likely to finish high school, and *less* likely to commit minor crime, than non-Cherokee youth in the area.

Akee and colleagues were able to do two other important things. First, rather ingeniously, they established that what mattered for the beneficial effect of the casino scheme on a household was not how far it was geographically from the casino, which might have been the case had the mechanism for the behavioural changes been, say, meeting lots of morally improving outside role models who had come to the area to use the casino facilities. (No, I don't think that's very plausible either, but the good thing about science is that you can try to test these possibilities against the data.) No, what mattered for the beneficial effect was just *how much money* came in to the household (you got more over time, remember, and the family got more the more registered Cherokee persons there were living in the household). The other thing the researchers were able to show was that a big part of the beneficial effect operated through creating better relationships within the household. The parents did not work any fewer hours as the free money increased (there's one for the Universal Basic Income advocates). As their financial situations improved, though, they reported higher quality relationships with one another and with their children. And the harmony was not

---

4   Akee, R. K.Q. et al. (2010). Parents' income and children's outcomes: A quasi-experiment. *American Economic Journal: Applied Economics* 2: 86–115, https://doi.org/10.1257/app.2.1.86

achieved by trading in their feckless spouses for new models, either: they just got on better with whoever they were already with. This makes sense: put people under less strain, and it's easier for them to get along well. A big way of taking the strain off is through the pocket book.

Akee, Costello and colleagues have one further set of results worth highlighting. They recently delved back into the questionnaires and evaluations supplied by the parents and, for some variables, the children, of Cherokee and non-Cherokee families.[5] They found, confirming previous analyses, that receiving the cash payments reduced symptoms of emotional disorders (basically anxiety and depression), and of behavioural disorders (basically being antisocial). Moreover, the researchers measured three of the 'Big Five' personality traits, Agreeableness, Conscientiousness, and Neuroticism. The former two are important for how you get on in life: Agreeableness describes the tendency to be cooperative and get along well with others, whilst Conscientiousness describes the propensity to be hard-working and organized. The classic successful bourgeouis is pretty Agreeable, and highly Conscientious. The prisons are full of people who rate low on both traits. And guess what: the arrival of income payments was associated with large increases in Agreeableness and Conscientiousness amongst young Cherokee. No comparable personality changes over time were seen in the non-Cherokee members of the study, who were living through the same general social period, but not getting the income increase.

In summary, then, the Eastern Cherokee casino income was 'helicopter money', a large increase in income that descends from on high with no skills training, no family counselling, no conditionality, and no prior logic. The CSJ hypothesis predicts that its arrival shouldn't have much improved the whole network of social problems. But by looking at how the Eastern Cherokee compare in social outcomes to their non-Cherokee neighbours both before the money arrives, and after, we can make pretty clean causal inferences about what raising incomes does. The evidence tells us unambiguously: relationships in families improve, kids stay on in school, kids become less likely to get

---

5   Akee, R. K.Q. et al. (2016). How does household income affect child personality traits and behaviors? *NBER Working Paper* No. 21562, https://doi.org/10.3386/ w21562

involved in minor crime and antisocial behaviour, addiction goes down, and even, most remarkable of all, *people's personalities change*. Not bad for treating the symptoms rather than the cause, eh?

§

Our second example comes from the developing world, from Kenya's Cash Transfer for Orphans and Vulnerable Children programme. This programme was a response to the fact that, due among other things to the AIDS epidemic, Kenya had a huge number of young people whose parents had died or were dying, and these young people needed supporting. Kenya could have spent its money in various ways: skills-training programmes, counselling, orphanages, and so on. It chose another path, a completely unconditional regular cash payment to the household in which the young orphan was living: helicopter money. So again, we have a nice clean test of whether cash alone does much for poor people with a manifold of different social problems. And better still, we have a proper randomised control trial of the programme. It was impossible to roll the programme out simultaneously in all of Kenya. Thus, districts were randomised to receive the programme immediately (the experimental group), or in a later wave (these districts served as the control group at the time of the evaluation, when their orphans had not yet received anything).

Comparing the two groups showed that the cash transfer improved school attendance rates, particularly when school was costly to attend (e.g. when the school was far away); and particularly when the children were older (which is the time when there is an opportunity cost of going to school, instead of generating money directly or looking after the household).[6] This is an important finding given that there was no conditionality in the programme whatever: the cash would continue to appear regardless of whether the young person stuck with school or not.

What I want to focus on, though, is a nice, revealing study of household expenditure in the control and the experimental groups,

---

6   Kenya CT-OVC Evaluation Team. (2012). The impact of Kenya's Cash Transfer for Orphans and Vulnerable Children on human capital. *Journal of Development Effectiveness* 4: 37–41, https://doi.org/10.1080/19439342.2011.653578

before and after the programme.[7] Experimental households spent more money on nearly everything once the cash started to roll in. This should not surprise us: after all, they had more money to spend. Here the researchers were able to do some clever econometric stuff, though. They used the pre-intervention spending data to construct a model of how household expenditures of different types in this population scale with income. This allowed them to make predictions: if this household behaved like a typical household in this population, then when you increase its income by 1500 shillings a month, how much more should we expect it to spend on food, how much more on healthcare, how much more on alcohol, etc.? The actual observed changes when the programme kicked in could then be compared to these predictions. In effect, the researchers asked, does the programme allow households to satisfy more completely the priorities they had anyway, or does it change their priorities?

When they performed this comparison, the researchers found that a number of categories of expenditure went up by less than expected, for example food. Within foods, expenditures on cheap tubers went down relative to expectation; it was only spending on high-quality foods that went up. Expenditures on alcohol and tobacco actually went down. Expenditure on healthcare went up relative to expectation, and households also saved and invested more than when the programme began. In short, as the cash landed, households shifted their preferences away from hedonic gratification (alcohol and tobacco) and immediate subsistence (tubers), and towards looking after their long-term health, and making investments. This is the riposte to the CSJ 'won't it be bad to just give poor people money when they don't have the skills to know how to spend it wisely' type of argument. They do seem to spend it wisely. Perhaps they are smart, and can figure out how to do so for themselves. Perhaps they are as smart as you, me or a development expert, but have had worse luck until now.

Both the Kenya example and the Cherokee one bring to light a very interesting conundrum: when you give people more money, their expenditure on narcotic substances goes down. Given abundant

---

7   Kenya CT-OVC Evaluation Team. (2012). The impact of the Kenya Cash Transfer Program for Orphans and Vulnerable Children on household spending. *Journal of Development Effectiveness* 4: 9–37, https://doi.org/10.1080/19439342.2011.653980

resources, it seems, most people don't value these things very highly. But, if people don't value them very highly, then why, when money was short (i.e. before the cash helicopter landed), were they spending anything on them at all in the first place? If something has a low value, then surely it would get crowded out when money was tight, and only perhaps creep in when the money supply gets looser? This conundrum really gets to the heart of the matter. The CSJ view looks at the behaviours of poor people, such as their proneness to use narcotics, and sees a disposition. It then says: shovelling cash on to this disposition won't do any good, and may even do harm. It's the disposition, stupid. The cash-first view, on the contrary, looks at the behaviours of poor people and says: that's a response to a situation. Change the situation (add money), and all kinds of decisions will follow suit.

This brings to mind classic animal research on addictive substances. Rats or mice living alone in small barren cages will self-administer morphine or cocaine enthusiastically, if given the chance. It turns out that the very same animals living in spacious, enriched stimulating environments, will do so significantly less, even when the drug is easily available.[8] In other words, the motivation to use rewarding narcotics is not a biologically inflexible drive in these creatures; it's a way of coping with adverse environmental contexts, the lack of alternative sources of reward, and it spontaneously though not completely fades away as those contexts improve.

This doesn't completely deal with the argument, in the human case, that goes as follows: why can't poor people just spend less on alcohol and tobacco in the first place? If they did so, it would be as if they were giving themselves a cash transfer programme. They could start to climb the ladder towards a better life using the money they saved, without having to wait for a casino to come along. This is a hard argument, but I expect an answer might be along the following lines. When you are poor, you occasionally do find yourself with a little left over, but this

---

8   Alexander, B. K., R. B. Coambs and P. F. Hadaway. (1978). The effect of housing and gender on morphine self-administration in rats. *Psychopharmacology* 58: 175–9, https://doi.org/10.1007/bf00426903; Chauvet, C. et al. (2009). Environmental enrichment reduces cocaine seeking and reinstatement induced by cues and stress but not by cocaine. *Neuropsychopharmacology* 34: 2767–78, https://doi.org/10.1038/npp.2009.127; Solinas, M. et al. (2009). Reversal of cocaine addiction by environmental enrichment. *Neuropsychopharmacology* 34: 1102–11.

surplus from the requirements of subsistence is small and temporally unpredictable. Thus, you can't plan to use it in any calculated escape route from poverty. When it does come, quite understandably, you treat it as a break in the weather, a small moment to medicate yourself from the awfulness and difficulty of life. But what poor people do when they get a small and unpredictable surplus in an otherwise bleak existence is not indicative of what they would do if we gave them a large and predictable permanent surplus. Both Kenya and the Cherokee case show us this. When you give poor people large, reliable, long-term surpluses, they start to behave just like people who have been lucky enough to have large, reliable, long-term surpluses in the first place. They don't need teaching; they don't need conditionality or monitoring. They just need the money.

§

The evidence we have reviewed has implications for two sets of things I care about: politics, and our view of human nature. First, the politics. Many cynical commentators suspected that the CSJ's de-emphasis of the purely financial aspects of poverty was a smokescreen for regressive policies. After all, if the bad thing about poverty is not the lack of money, then there is no compelling case for income redistribution. The proposal to measure poverty using non-income indicators makes it much easier to enact financially regressive policies and not have anyone notice that is what you are doing. In support of this point, the tax and benefit changes enacted by the Cameron government and its successor between 2010 and now have clearly been regressive: the households in the lower deciles of the income distribution have seen their incomes eroded much more sharply (even in absolute terms, never mind proportional ones) than those at the top.[9] So this is what the promised assault on poverty actually looked like: redefine poverty as being not about money, and then take money away from poor people. There is, I suppose, a kind of logic to it.

In my view, the political implications of cases like the Eastern band of Cherokee are clear: we need to redistribute income and wealth more,

---

9   Equality and Human Rights Commission. (2018). *The cumulative impact of tax and welfare reforms.* Downloaded from https://www.equalityhumanrights.com/en/ publication-download/cumulative-impact-tax-and-welfare-reforms

from the people at the top of the distribution to whom it brings little benefit, to the people at the bottom for whom, as we have seen, it makes a profound difference.[10] So impressive are the Cherokee results that redistribution becomes not just a moral obligation, but a simple matter of pragmatics. A thought experiment: Say I tell you I've developed an intervention programme that costs a bit, but has been shown to reduce crime, increase educational attainment, reduce drug and alcohol dependency, promote family stability, and make people nicer and more conscientious. That sounds great you say, perhaps imagining fish oil supplementation, or mindfulness-based cognitive therapy. That intervention should be funded nationally. Now I tell you what my intervention is: it's called making sure people aren't poor. You might suddenly feel less keen, even if I tell you what could well be true: it's cheaper and more effective than the alternatives in the long run.

Now, what about our view of human nature? The evidence from cash transfers strikes quite strongly against any view in which individual differences in behaviour are the result of some fixed inner essence that is obtained early in life and inflexible thereafter (disposition, taste, personality, culture, or whatever you want to call it). If such fixed-essence views were correct, then the CSJ would be basically right: helicoptering cash to people would just lead them to perpetuate whatever they are doing already, but under looser constraints. The alternative to the fixed-essence view places situation and context more centrally as drivers of people's immediate behaviour; sees people as highly plastic in their tastes, strategies and decisions; and emphasises that a big part of where we end up in life is due to (reversible) luck.

One influential version of the fixed-essence view has genetic differences doing a lot of the work. You might think that your personality, for example, is largely due to your genetic inheritance.[11] If your lineage is disagreeable and not conscientious, it drifts down the social ladder. If you are lucky enough to have agreeable and conscientious genes, you climb up. But this cannot be the whole story: One of the reasons I love the Cherokee example so much is that it shows personality, that most fixed- and genetic-feeling of things, changes in response to helicopter

---

10   See Why inequality is bad, this volume.

11   The evidence is reviewed in Nettle, D. (2007). *Personality: What Makes You the Way You Are* (Oxford: Oxford University Press).

money. So what is going on? Are psychologists wrong about personality being fixed and heritable?

They are not entirely wrong, but some care is needed. There seem to be heritable influences on personality, as evidenced from studies of twins; and two people facing the same environment of poverty can respond to it quite differently, which might well have something to do with dispositional differences between them. But it is too great a leap to move from 'genetic differences explain some of the variation in personality between individuals facing similar environments' to 'group differences in personality are best explained by variation in genes between those groups'.

Let me propose the following analogy. Imagine you grow corn on a field uniformly rich in fertilizer. All of your corn plants will be tall, but some of them will be a little taller than others. The differences in height between your plants will probably be mainly due to genetic variation between them (after all, they all developed in the same benign environment). If you did a study of your plants at this point, you would conclude that variation in height is highly heritable, mainly a matter of genetics. Now you take away the fertilizer from half the field. The plants in that half of the field grow much less tall in the next year. This is entirely caused by the change in inputs. So although you had concluded that the individual differences in plant height were heritable when the environment was good, you have also proved that a group difference (between one half of the field and the other) has nothing to do with genetics, and everything to do with environmental factors. And so, I think, with people: there may well be genetic variations, but where we really see their importance is in explaining the residual variation given a constant environmental context. The environmental context for different social groups is nowhere near constant, though, and that's a much more relevant explanatory principle for differences in average level across groups.

Another form of fixed-essence thinking puts culture in the place of genes. The argument here is that culture is a pseudo-genetic inheritance system: you absorb a system of behaviours via social learning in your childhood, and thereafter you are pretty much stuck with it. Social change, when it happens, is a matter of cultural mutations that gradually change in frequency over the course of generations, faster than genetic change, perhaps, but still slower than the individual lifetime.[12] This kind of thinking can't really explain the Cherokee case. The cash changed behaviour patterns massively within a single cultural generation, and without (as far as we know) changing who learned their culture from whom. Whatever people were doing, it was not internalising and persisting with the behaviours they had been exposed to in childhood. Instead, you have to see people as strategic agents who change their decisions and dispositions more or less in real time as their environments (and their information about their environments) change. This does not preclude roles for norms and social transmission in an account of human behaviour; but it does warn us against pushing the analogy between cultural and genetic inheritance too far.[13]

§

In spite of everything, I still find it hard to accept that the best thing I could do to help poor people is just to give them my money. I know that many other people feel the same. At the bottom of this, I think, is some kind of *illusion of the validity of our expertise*. The idea of an illusion of validity comes from classic work by Daniel Kahneman and Amos Tversky.[14] Kahneman served in the psychology branch of the Israeli army. This branch developed expert methods for predicting who would make a good officer. Kahneman followed up and found that the expert methods for officer selection were, in fact, quite useless: you might just as well have chosen every third recruit and stuck a badge on them. Nonetheless, the psychology-branch experts remained convinced that their expertise was valid, and continued self-importantly to deploy it. As well as over-valuing our own expertise, I think we are all prone to under-value the expertise of people we consider unlike us, in this instance, poor people.

We seem to feel sure that we have a good analysis of how poor people could make their lives better, so sure that we are not shy of

---

12   This view is particularly clearly articulated by Mathew, S. and C. Perreault. (2015). Behavioural variation in 172 small-scale societies indicates that social learning is the main mode of human adaptation. *Proceedings of the Royal Society*, B. 282: 20150061, https://doi.org/10.1098/rspb.2015.0061

13   See The cultural and the agentic, this volume.

14   Kahneman, D. and A. Tversky. (1973). On the psychology of prediction. *Psychological Review* 80: 237–51, https://doi.org/10.1037/h0034747

coming up with advice, diagnoses, intervention strategies, and training programmes. On average, these may well do less good than just transferring the equivalent amount of money to low-income households. That's hard to accept, especially for a professional academic whose day job is having something expert-sounding to say. It's hard to accept because conceding the value of just transferring money is tantamount to admitting that my expertise in how to fix things is low. I might think I am brilliant, but in truth I would probably be really bad at living on the bread-line: I have not developed the skills. Poor people, on the other hand, are generally going to be more expert at coping with that context. Therefore, on average, they are probably going to make better decisions about how to navigate the shoals than I am. Cash transfer takes micro-allocation decisions out of the hands of people who don't really know what they are doing (like me), and into the hands of people with expertise (the recipients). Accepting the case for cash transfer then, is really about accepting that poor people are cognitively equivalent to rich people, but on average more skilled; and therefore trusting them to make their own decisions. To do this requires letting go of the intuitions that give us paternalism and the idea that poor people are deficient in decision-making capacity. And those intuitions, I suspect, are more deeply embedded, even amongst progressive academics, than it is comfortable to admit.

10. Getting your head around the Universal Basic Income

> Can we not find a method of combining [the advantages of anarchism and socialism]? It seems to me that we can [...]. The plan we are advocating amounts essentially to this: that a certain small income, sufficient for necessaries, should be secured to all, whether they work or not
>
> –Bertrand Russell[1]

> A host of positive *psychological* changes inevitably will result from widespread economic *security*.
>
> –Martin Luther King[2]

Today should be the best time ever to be alive. Thanks to many decades of increasing productive efficiency, the real resources available to enable us to do the things we value—the avocados, the bicycles, the musical instruments, the bricks and glass—are more abundant and of better quality than ever. Thus, at least in the industrialised world, we should be living in the Age of Aquarius, the age where the most urgent problem is self-actualisation, not mere subsistence: not 'How can we live?', but 'How shall we live?'.

Why then, does it not feel like the best time ever? Contrary to the predictions of mid-twentieth-century economists, the age of universal wellbeing has not really materialised. Working hours are as high as they were for our parents, if not higher, and the quality of work is no better for most people. Many people work several jobs they do not

---

1   Russell, B. (1918). *Roads to Freedom. Socialism, Anarchism and Syndicalism* (London: Unwin Books, p. 80–1).
2   King, M. L. (1968). *Where Do We Go from Here: Chaos or Community?* (Boston: Beacon Press, p. 173).

     https://doi.org/10.11647/OBP.0155.10

enjoy, just to keep a roof over their heads, food on the table, and the lights on. In fact, many people are unable to satisfy these basic wants despite being in work: the greater part of the UK welfare bill, leaving aside retirement pensions, is spent on supporting people who have jobs, not the unemployed. Thousands of people sleep on the streets of Britain every night. Personal debt is at unprecedented levels. Many people feel too harried to even think about self-actualisation.

Twin spectres stalk the land, and help explain the gap between what our grandparents hoped for and what has materialised. These are the spectres of inequality and insecurity. Insecurity, in this context, means not being able to be sure that one will be able to meet one's basic needs at some point in the future, either because cost may go up, or income may fluctuate. Insecurity is psychologically damaging: most typologies put security as one of the most basic human emotional needs.[3] Insecurity dampens entrepreneurial activity: one of the big reasons that people don't follow up their innovative ideas is that these are by definition risky, and they worry about keeping bread on the table whilst they try them out. Insecurity deters people from investing in increasing their skills: what if they cannot eat before the investment starts to pay off? It encourages rational short-termism: who would improve a house or a neighbourhood that might be taken away from them in a few months' time for reasons beyond their control? It also increases the likelihood of anti-social behaviour: I would not steal a loaf of bread if I knew there was no danger of going hungry anyway, but faced with the danger of starvation tomorrow, I would seriously consider it.[4] Insecurity is a problem that affects those who have little to start with especially acutely: hence the link between insecurity and inequality.

Big problems require big ideas. Our current generation of politicians don't really have ideas big enough to deal with the problems of widespread insecurity and marked inequality. Big ideas come along every few decades. The last one was about forty years ago: neoliberalism, the

---

3   Maslow, A. H. (1943). A theory of human motivation. *Psychological Review* 50: 370–96, https://doi.org/10.1037/h0054346; Griffin, J. and I. Tyrrell (2003). *Human Givens: A New Approach to Emotional Health and Clear Thinking* (Chalvington, East Sussex: HG Publishing).

4   A point made by Thomas More, in his Utopia, as long ago as 1516: "[…] no penalty on earth will stop people from stealing, if it's their only way of getting food". Presciently, More goes on to suggest that "[providing] everyone with some means of livelihood" is thus a way to deal with the problem of petty theft.

idea that market competition between private-sector corporations would deliver the social outcomes we all wanted, as long as government got out of the way as far as possible. Interestingly, neoliberalism was not such an obviously good idea that politicians of all stripes 'just got it'. It took several decades of carefully orchestrated deliberate communication and advocacy, which was not at all successful at first, to eventually make it seem, across the political spectrum, that the idea was so commonsensical as to be obvious.[5] I don't think any of the early advocates of neoliberalism could possibly have dreamed that after thirty years of implementation of their big idea, available incomes would have stagnated or declined for the median family; public faith in corporate capitalism would have seeped away; even the UK Conservative party would have to concede that market mechanisms did not really work as envisaged;[6] or that the major UK political parties would both be advocating government-imposed price-caps in an area, the supply of energy, where the neoliberal market model had been followed to its logical conclusion. It feels like we are washed up on the end of one big idea, waiting for something else to come along.

Our current politicians propose to deal with symptoms piecemeal—a minimum-wage increase here, a price cap there, rent-control in the other place; tax credits for those people; financial aid to buy a house for those others. At best we are dealing with one symptom at a time. Each piecemeal intervention increases the complexity of the state; divides citizens down into finer and finer *ad hoc* groups each eligible for different transactions; requires more bureaucratic monitoring; and often has unintended and perverse knock-on effects. For example, helping young people to buy a house with government financial aid only maintains the high levels of house prices. Vendors can simply factor into the price the transfer from government that they will receive. The policy would be much less popular if millions of pounds of taxpayer money were just given directly to large property development corporations, but that might as well be what the policy did. No, something more systemic is needed; an idea with bigger and bolder scope. That big, bold idea just might be the Universal Basic Income.

§

---

5   See Bregman, R. (2017). *Utopia for Realists: And How We Can Get There* (London: Bloomsbury) on this point.

6   For example: "We do not believe in untrammelled free markets", *Conservative and Unionist Party Manifesto, 2017 general election*, p. 9. Downloaded from: https://www. conservatives.com/manifesto

A Universal Basic Income (UBI) is a regular financial payment made to all eligible adults, whether they work or not, regardless of their other means, and without any conditionality whatever. Receiving it is a fundamental entitlement that comes with being a member of society: people can know that it will always be there, now and in the future. It should not be a fortune, but it should ideally be enough that no-one ever needs to be hungry or cold.

The arguments for the UBI are well summarised elsewhere, and so I will not repeat them at length.[7] All developed societies agree on the need to protect citizens from desperate want that may befall them, usually for reasons beyond their control. However, the ways we currently make these transfers are incredibly complex. Guy Standing reports that in the USA, there are at least 126 different federal assistance schemes, not to mention state-level ones.[8] In the UK, individuals have had until recently to be separately assessed for unemployment support, ill-health support, carer support, working tax credits (which amount to low-income support), and so on. The new Universal Credit system only partly simplifies this thicket. Each conditional scheme generates a bureaucracy of assessment and the need for constant eligibility monitoring, at vast expense.

Moreover, conditional transfers always generate incentive problems. If you go back into work after being unemployed, you lose benefits. If you are a carer and the person you care for recovers, you are financially penalised: you do better by keeping them ill. If your wages or hours go up, you lose out in benefit reductions. Under the UK's new Universal Credit system, the marginal tax rate (the amount you lose of every extra pound you earn in the job market if you are a recipient) is around 80%, and that scheme was a reform designed to increase the incentive to work! Moreover, the 80% figure does not factor in the fact that if you move briefly out of eligibility, for example for some seasonal work, you are uncertain about when and whether you would be able to get back in afterwards, should you need to. This is a disincentive for taking the work. It is very hard to eliminate these perversities within any system of conditional, circumstance-specific transfers.

---

7   I recommend in particular Standing, G. (2017). *Basic Income: And How We Can Make It Happen* (London: Penguin), which also provides a history of the idea.

8   Standing, G. (2017). *Basic Income: And How We Can Make it Happen* (London: Penguin, p. 53).

The UBI, then, seems like a good idea. It is far from a new one. It has fragmentary roots in the eighteenth and nineteenth centuries. In the twentieth century, there was one wave of enthusiasm in the 1920s, and another in the late 1960s and 1970s. The second wave generated a positive consensus, specific policy proposals, and a certain amount of pilot activity, but other paths ended up being taken. The idea has never quite died, though. It is now back in political consciousness in a very big way.

§

Why, when the UBI seems such a good idea, when it has been cognitively available to us for so long, when so many very clever people have modelled it and found it desirable, is there no developed society on earth in which it has been fully implemented? Partly this is because democratic governments, indeed societies in general, are poor at far-reaching systemic reform, instead finding it easier to tinker with and tune existing systems. It's only the political outsiders who dare propose massive change—they have less to lose. But it is also because human psychology is an obstacle to the UBI, and this is what interests me in this essay. As Pascal Boyer and Michael Bang Petersen have recently argued,[9] when we (non-specialists) think about how the economy ought to be organized, we don't derive our conclusions from formal theory, simulations, or systematic research evidence. No, we generally fall back on simple social heuristics, like 'if someone takes a benefit, they ought to pay a commensurate cost'; 'more for you is less for me'; or 'people should only get help when they are in need'. These simple social heuristics are all well and good for the problems they developed to solve—basically, regulating everyday dyadic or small-group social interactions. But they don't automatically lead us to the right conclusions when trying to design optimal institutions for a complex system like a modern capitalist economy.

Certain aspects of the UBI idea violate one of these simple social heuristics. In fact, the UBI sometimes manages to violate two different and contradictory simple social heuristics simultaneously, as we shall see. These violations are like notes played slightly out of tune: they just

---

9   Boyer, P. and M. Bang Petersen. (2018). Folk-economic beliefs: An evolutionary cognitive model. *Behavioral and Brain Sciences* 41: e158. https://doi.org/10.1017/S0140525X17001960.

seem wrong, before one has had to think much about it. Politicians are afraid of these reactions; they don't like going out to campaign and meeting the same immediate objections all the time. If you want to build a consensus for the UBI, you have to analyse these jarring notes with some care, and develop a counter-strategy. For UBI to go mainstream, a positive case will need to be made that *also* draws on easily-available simple social heuristics. If we can't make it make intuitive sense, it will be confined forever to the world of policy nerds.

Fortunately, the challenge can be met. Our simple social heuristics do not constitute a formally consistent system, like arithmetic (why would they?). Instead, they are a diverse bunch of often contradictory gut feelings and moral reactions each triggered by particular contextual cues. For example, we do have strong intuitions that people should not take a benefit without paying a commensurate cost, but these intuitions only get triggered when certain sets of features are present in the situation. These features include: the resource is scarce enough every additional unit of it is valuable to me; the resource was created by deliberate individual effort; the person taking the benefit is somehow dissimilar to me, so their interests are not closely tied in to mine; and it is feasible to monitor who is getting what at reasonable cost. The features do not always obtain: the resource might be more plentiful than anyone really needs; its acquisition might be mainly due to luck; the other people might be fundamentally similar to me, or their interests closely bound up with mine; or the cost of monitoring who got what might be prohibitive. In such situations, humans everywhere merrily and intuitively sign up to the proposition: the resource should be shared out somehow. There are a number of ways this can happen: pure *communal sharing*, where each qualifying individual just takes what they like, or *equality matching*, where every qualifying individual is allotted an equal share as of right. Every society has domains in which communal sharing or equality matching is deployed in preference to market pricing (the rule 'you should only take a benefit if you pay a commensurate cost').[10]

Hunter-gatherers deal with large game—chancy and producing a huge surfeit when it comes—by communal sharing. Even in the more private-property focussed Western societies, communal sharing is

---

10   See Fiske, A. P. (1991). *The Structures of Social Life* (New York: The Free Press); and Rai, T. S. and A. P. Fiske. (2011). Moral psychology is relationship regulation: Moral motives for unity, hierarchy, equality, and proportionality. *Psychological Review* 118: 57–75, https://doi.org/10.1037/a0021867

ubiquitous. Households, for example. If I buy a litre of milk, I don't give my wife a bill at the end of the week for whatever she uses. *Su casa es mi casa.* Communal sharing or equality matching happens beyond households too. It is anathema to suggest that the residents of Summerhill Square might charge passersby for the air they breathe whilst walking through. Very few people think that those who pay more taxes should get more votes. When proposals are made to move a resource from the domain of the communally shared or equality-matched to the priced, there is outcry: witness the response that greets proposals for road tolls in places where use of the roads is currently free; or to charge money at the gates of the town park. The case for the UBI is the case for moving part—no means all—of our money the other way, out of conditionality and into the domain of the equality-matched. Getting your head around it involves framing your understanding of our current economic situation in such a way as to trigger the appropriate equality-matching intuitions. Here as in many other political domains, those who determine the framing of the problem get to have a big influence on the outcome.[11]

§

Whenever one talks about the UBI, one hears the same objections, including:

1. How can we afford such a scheme?

2. Why should I give my money to people for them to do nothing in return?

3. Why would anyone work if they were given money for free?

4. Why should we give money to the rich, who don't need it?

The first of these objections is the easiest to dispose of. There have been detailed recent costings for the UK, which vary in their assumptions, but the consensus is that introduction of a modest initial UBI scheme would require surprisingly little disruption to our current tax and expenditure

---

11 Elcheroth, G., W. Doise and S. Reicher. (2011). On the knowledge of politics and the politics of knowledge: How a social representations approach helps us rethink the subject of political psychology. *Political Psychology* 32: 729–58, https://doi. org/10.1111/j.1467-9221.2011.00834.x

system; perhaps modest tax rises, perhaps no change, perhaps tax cuts.[12] If this surprises you, let me give you the following back-of-an-envelope calculations. There are around 65 million people in the UK, of whom 63% are aged between 16 and 64. Assuming that the over 65s will continue with their current pension arrangements instead of the UBI, that gives us at most 41 million adults to cater for, plus about 12 million under-16s. Let's say we want to give £80 per week to each of the adults. This would cost £171 billion per annum. And let's further say that we want to give £40 per week, to the mother or other caregiver, for each child under 16. That's another £25 billion, giving a nice round £200 billion in total.

Of course, £200 billion a year is an eye-watering sum. But UK government expenditure in 2017 was £814 billion,[13] so we are only talking about one quarter of what the government spends anyway. Increasing government expenditure by one quarter might be a rather rash move, but this would not be the net increase, because the UBI would produce savings elsewhere. The welfare bill for 2017, less retirement pensions, was £153 billion.[14] It's unrealistic to expect a UBI scheme to reduce this to zero: most UBI advocates argue for retaining some extra provision for the disabled, and also retaining, for the time being, means-tested benefits to pay housing rental in some cases (the cost of housing is so high in parts of the UK that many people would become homeless if this disappeared overnight). But certainly, we might hope to eliminate up to £100 billion, or 2/3, of the non-pensions welfare bill, including a very large part of the administrative cost. So we are already half-way there.

At present, most UK adults are taxed at a zero rate on the first £8,164 of earned income, 12% from £8,164 to £11,500, and 32% above £11,500. What this means, in effect, is that anyone earning £11,500 or more is effectively being given a freebie from the state of £3680, compared to being standardly taxed at 32% from the first pound. This figure—£3680 per year—is, you will note, not so very far off my proposed initial UBI of £4160 anyway. Personal tax allowances cost the government around

---

12 Torry, M. (2016). An evaluation of a strictly revenue neutral Citizen's Income scheme. *Euromod Working Paper Series* EM5/16. Downloaded from: https://www.iser.essex.ac.uk/research/publications/working-papers/euromod/em5–16; Painter, A. and C. Thoung. (2015). *Creative Citizen, Creative State: The Principled and Pragmatic Case for a Universal Basic Income.* (London: Royal Society of Arts).

13 Information from: https://www.ukpublicspending.co.uk/total

14 Information from: https://visual.ons.gov.uk/welfare-spending/

£100 billion per annum in foregone revenue.[15] If my proposed UBI were to be introduced, it would be reasonable to ask people to pay their taxes from the first pound. For people like me who earn more than £11,500 per annum, the introduction of the UBI would then be largely neutral, my tax bill going up by around £4000, offset by £4000 coming separately into my bank account as UBI. So, if you will allow me very broad approximations, moving to a modest UBI would cost about £200 billion per annum, to be funded by about £100 billion of welfare savings, and about £100 billion from abolishing personal tax allowances—so pretty much fiscally neutral. And this is just a business-as-usual analysis of the likely financial consequences. What advocates believe is that there will be positive knock-on effects: people will be able to move to more productive and enjoyable jobs, or start entrepreneurial activities; people have no financial disincentives to take casual work or increase their hours; the expensive negative psychological consequences of insecurity (anxiety, depression, addiction, maybe even crime) will improve. Thus, what you end up with will be a net saving for the government, not a net cost.

The initial scheme discussed above, and other proposals like it, are not immediately very redistributive. Those currently receiving full Universal Credit would only end up with about the same as their current entitlement; and, as I mentioned above, for well-off people like me, the UBI would be almost exactly offset by the increase in my tax bill.[16] So what is the point of such a reform? The answer has to do with security. I see UBI not so much as an immediate solution to inequality (you would have to set it very high to have a big direct effect on the inequality figures), but as a prophylactic against insecurity. For a wealthy person such as myself, there's not much financial difference between getting a personal tax allowance and receiving a UBI, *until* my life is hit with a shock. I am well-off now, but I might not always be. Say I suddenly lose my job, or need to care for my wife. I know the UBI will continue to be there, every week, without any action required of my part. I can factor it into my worst expectations. The same is not true of the transfer effected by my personal tax allowance. And this, briefly, is the best response to objection 4, 'Why

15 Standing, G. (2017). *Basic Income: And How We Can Make It Happen* (London: Penguin, p. 131).

16 The group that would clearly benefit in income terms from a scheme such as this one is those just well enough off to lose conditional benefit entitlements, but still financially constrained: the 'squeezed middle'.

should we give money to the rich, who don't need it?'. Well, as long as they remain rich, then they are net payers into the system, since their tax bill exceeds their UBI, so we are giving them money only in an accounting sense. But it is still better to have them make a large tax payment in and concurrently take a small UBI payment out, rather than just make their tax rate a bit lower, because they might suddenly become non-rich at any moment. The UBI is ready for that moment should it come. To counter objection 4, we need to activate the social heuristics: 'anyone could have bad luck' and 'everyone is potentially in the same boat'.

There is a large difference between the knowledge that £80 a week will always come into my bank account, this week, next month, and for the rest of my life; and the knowledge that, if things go badly for me, I can do a complex application process, be subjected to a humiliating and lengthy bureaucratic examination, following which, after a delay of up to six weeks during which I will receive nothing, about £80 per week may or may not start to appear in my bank account, and could be withdrawn at any moment if I am ten minutes late for an interview, or am deemed not be sick enough or not be trying hard enough to look for work.[17] It is ironic that the system we often refer to as 'social security' provides the exact opposite of that: it provides continual, unplannable-for uncertainty akin to a sword of Damacles. The insecure, such as those waiting for benefits decisions or enduring benefits sanctions, have short-term problems of liquidity. They lose their homes and possessions, or end up having to borrow money at very high interest rates. This is expensive and spirals them into abject poverty. Reducing insecurity could have an indirect effect on inequality, by stopping this spiral. And the health and wellbeing benefits observed in trials of UBI and minimum income guarantees, even over quite short periods, have been so massive that it is hard not to conclude that security does something interesting to human beings, out of all proportion to the monetary value of the transfer, just as Martin Luther King predicted.[18]

---

17 One in five Universal Credit applications is rejected because of some procedural error, leading to many weeks with no income. See 'Complex rules for universal credit see one in five claims fail', The Guardian, May 12th 2018, https://www.theguardian.com/society/2018/may/12/one-in-five--turned-down-for-universal-credit-rules-too-complex

18 See inter alia: Widerquist, K. (2005). A failure to communicate: What (if anything) can we learn from the negative income tax experiments?. *Journal of Socio-Economics* 34: 49–81, https://doi.org/10.1016/j.socec.2004.09.050; Forget, E. L. (2011). The town with no poverty: The health effects of a Canadian guaranteed annual income field experiment. *Canadian Public Policy* 37: 283–305, https://doi.org/10.3138/cpp.37.3.283;

§

What about objection 2 ('Why should I give my money to people for them to do nothing in return?'). The objection has two parts: there's a part about my money being *my* money, and a part about *giving to other people without them doing anything in return*. Both parts are important.

First, the *my money* part. All societies distinguish between individually-owned resources and communal resources, though they draw the line in different places. Across societies, alienating an individually-owned resource from someone is morally wrong; but depriving people of a communal resource is equally so. The kinds of cues that trigger intuitions of individual ownership are: my having transformed the material extensively through deliberate action; the resource having been given to me by someone in return for something specific; or the resource having been in my sole possession and use for some time. The kinds of cues that trigger intuitions of communal ownership are: the resource being very abundant; its use being hard to monitor and police; a little of it being essential for everyone's survival; and the having of it being mainly due to luck. So I think a first move you need to make in making the UBI make sense is to loosen the hold of the individual ownership schema on the money in your wage packet.

The money in my wage packet certainly feels like a good candidate for individual ownership. I have worked hard to get where I have, and this leads to the intuition that every penny in my wage packet is mine, should not be given away to other people without a specific reciprocal service rendered. I supposed I should grudgingly admit that I have got *some* help from others in earning my salary as an academic—I mean it's not *quite* all my own sweat. Following the logic of individual ownership, I should really have paid for all these inputs at point of use, but somehow I didn't always do so. There's the statistical computing language R, the backbone of all my research; developed by people I didn't know and made freely available without me lifting a finger. Maybe 1p in every pound I earn is really owable to the R Foundation for Statistical Computing. Then come to think of it there is the computer

Basic Income Grant Coalition, Namibia. (2008). *Basic Income Grant Pilot Project Assessment Report* Downloaded from http://www.bignam.org/Publications/BIG_Assessment_report_08a.pdf; and Standing, G. (2017). *Basic Income: And How We Can Make It Happen* (London: Penguin).

itself, developed by a mixture of public and private investment mainly before I was born. It's unthinkable that I could be a productive modern professor without this input available. So really I should attribute 2p of each pound I earn to having had that available. Come to think of it, I could not really earn anything as a professor without the existence of an affluent society in which enough people are freed from daily subsistence activities as to want to spend their time studying behavioural science. So I guess I owe the Industrial Revolution say 5p; and then another 3p to those Europeans who invented a rather good system of universities for students to come and study at. Oh, and I do use the scientific method rather a lot (say 4p distributed across a wide range of people in many countries over the last couple of hundred years, and another 2p specifically for the intellectual work of creating my discipline). And a couple of pence in the pound for the philosophers of the enlightenment; without them to make the world safe for my kind I would at best be a priest with low wages. And then there's the Romans. What did the Romans ever do for me? Well, there's the sanitation. And the roads....

As soon as we complete this exercise, we are forced to concede that what seems like *my money* only partly meets all the triggers for individual ownership (my individual labour produced it). In large part, it is a windfall of cumulative cultural evolution. I just got lucky to be born into a shared cultural and technological heritage. I can't pay back to all those parties whose cultural activities contributed to my luck, since many of them are long gone (and besides, they are innumerable and diverse). But accepting that what I earn is partly due to an abundant social windfall created by a whole society over time, whose use and scope is hard to monitor, and I acquired by sheer luck, loosens the hold of the intuition that all my money all belongs exclusively to me. It's a short step from 'a part of what I receive from society is due to our common, difficult to monitor, abundant social luck' to 'a part of what I receive should be shared out'.

So now we turn to the part about why I should give anything to strangers without requiring them to pay any particular cost in return. A popular pro-UBI argument here, which goes back to Thomas Paine, is that people should be recompensed for the natural heritage that has been alienated from them. The land has been enclosed and privatised; the water has been bottled and sold; you can't just chop down the trees, hunt game or build a house where you want, as you would have been able to do at the dawn of society. The UBI is this recompense—the royalty, if you will, on an inheritance that was once socially shared but has been taken away by civilization. This reasoning is fine, but a bit lofty and philosophical. I prefer a quiverful of different, more forward-looking arguments.

First, social transfers of some kind are necessary, and monitoring them under the current system is really costly. The UK government recently announced that it needed to review whether its rules on disability benefit claims had been applied correctly to recent claimants.[19] This review is estimated to cost £3.7 billion. That's enough to give my proposed UBI to everyone in the town of Hexham for over 8 years. Not the cost of the benefit, not the cost of administering the benefit, just the cost of one review of whether the benefit has in fact been correctly administered, for a benefit that only a small fraction of the UK population claims anyway. Scale that up and you appreciate the madness of how we currently administer social transfers.

Second, I *do* derive all kinds of payoffs from the welfare of others, even strangers. What are they? Well, I enjoy strolling around my city. I enjoy living in a nice orderly street. I enjoy going to the theatre. If my co-citizens were so hungry and desperate that they turned to assaulting their fellows, smashing property, not tending their yards, and abandoning the arts, my personal wellbeing would be directly reduced. I like writing books and giving lectures. It's therefore in my direct interest that as many people as possible have the resources to read or attend these. Businesses can only flourish if there are people well enough off to be customers. This was the great insight of Henry T. Ford: he realised he could really make a lot more money once he paid his workers enough that they would be able to buy his cars. It's the kind of reverse Ponzi-scheme trick, or perpetual motion machine, of modern consumer capitalism: those at the top of the pyramid need enough money to get down to those at the bottom of the pyramid that those people can buy goods and services, which means that the money

---

19   See 'Government to review 1.6m disability benefit claims after U-turn', *The Guardian*, January 29th 2018. https://www.theguardian.com/society/2018/jan/29/government-to-review-16m-disability-benefit-claims-after-u-turn

comes back up to them again. Otherwise the whole thing grinds to a nasty halt.

One way of thinking about this is to say that, in a community, because of the fundamentally social character of human life, the well-being of each individual creates a spill-over benefit for the others. It's what economists call a positive externality. Because of the changes in behaviour that will follow from my neighbour not being in completely dire straits, *my* life improves a tiny little bit as theirs does. This improvement is very real and substantial, but hard to tie to any one act my neighbour does, and hence hard to monitor or account for in a ledger.

Third, the marginal wellbeing returns to keeping all of my money are diminishing. Diminishing marginal returns mean that if the first few hundred pounds of income massively improve my well-being, then the next few hundred improve it slightly less, and so on. A few years ago, Karthik Panchanathan, Tage Rai, Alan Fiske and I produced a simple model of what resource distribution a selfish actor should prefer when there are positive social externalities, and diminishing wellbeing returns. We imagined a simple world where there are two actors, me and someone else. We put a value s on the positive externality that flows to me as the other person's well-being increases by one unit. Now we ask: if I can decide how all the available resources get divided up, what allocation should I prefer? The exact numerical answer depends on the value of s and the degree to which marginal returns diminish, but generally, the result is the following. I should want to keep everything up until the point where I myself have got off the steepest part of the increasing wellbeing curve. Above that, it becomes rational for me to want the *other* actor to have the next chunk of resource, since the positive social externality coming to me from their large increase in wellbeing (they are still on the steep bit of the curve, remember) outweighs the rather small increase in my wellbeing I get from keeping it (since I am on the flatter bit of the curve).[20] There is no 'problem of

---

20 Exactly, I should want the other actor to have the next unit of resource as long as sb > c, where s is the size of the social externality, b is the marginal wellbeing gain of the other actor, and c is the wellbeing gain I would receive by keeping the resource for myself. Nettle, D., K. Panchanathan, T. Rai and A. P. Fiske. (2011). The evolution of giving, sharing and lotteries. *Current Anthropology* 52: 747–56, https://doi.org/10.1086/661521

cheating' in this model, since we assume that the positive externalities arise from behavioural changes that the other party will simply want to make anyway as their state improves. It's a model of mutual benefit, or interdependence, rather than tit-for-tat.

This is the reasoning I would use with a well-off person to advocate funding a UBI from their taxes. The money you put into other people's UBIs will directly increase your individual wellbeing, because in a society where no-one is desperate, it's easier for the things you really value and derive benefit from to flourish. Furthermore, as already discussed, UBI offers security to you too. You may not need it right now, but you could do in the future. Both of these are self-interest arguments, where self-interest is construed sufficiently broadly. You have to be careful about basing all policy arguments on self-interest: it can end up signalling that self-interest is the only normal reason for action, which could become a self-fulfilling prophecy.[21] Nonetheless, perhaps here having self-interest on side helps buttress nobler motives. Experience shows that the long-term success of social policies is tied to the relatively well-off seeing themselves as getting something from them. Where schemes are perceived to benefit only an 'underclass', different in kind from the people footing the bill, support is easily driven away in the next downturn.

§

Objection 3 ('Why would anyone work if they were given money for free?') is based on the reasonable intuition that conditionality is important in motivating others to do something. One does not generally say to the plumber: 'Here's £100. I'm hoping that at some point you will fix my tap'. However nice the plumber might be, the incentives are a bit wrong here. And if people withdrew their supply of labour, the very affluence that can fund the UBI would be undermined.

The best way to loosen this objection is to remind one's interlocutor of two things. First, the UBI is only ever going to be basic, and people want more than basic out of life. If people's life ambitions were limited to gaining some modest level of income of £5000 or £10000 per annum a year and then stopping, then frankly, the behaviour of the vast majority

---

21 See Bowles, S. (2016). *The Moral Economy: Why Good Incentives Are No Substitute for Good Citizens* (Yale: Yale University Press) for discussion.

of people in Western societies for the last century would be completely incomprehensible. Lottery winners almost universally continue to work, though often not in their previous jobs. Academics don't work less when they become full professors: they work harder. The very same critics who say that people won't do anything if given money for free also often advocate the awarding of huge salaries — millions of pounds per annum — to CEOs and other leaders. Admittedly, those huge salaries are conditional on working, whereas the UBI is not. But the fact that the salary allegedly needs to be so huge to attract candidates implies that people are motivated not just by getting a little bit of money, but by getting a lot. So those who advocate large salaries must believe that the motivation for more money holds up at levels of income way above the basic (at least for the right sort of people, but hey, maybe all people are the right sort).

Second, more important than the *amount* of labour people supply is the *productivity* of that labour. By this, I mean people choosing to do activities that are socially useful, in which they are happy, and that they are good at. That has to be key to maximising social wellbeing as well as economic stability in future. There is plenty of evidence from pilot schemes of the effect of the UBI (or similar policies) on labour supply. In the 1970s North American schemes, reductions in work hours were real but very modest. No-one stopped working altogether (and these were minimum income guarantee schemes, which provide stronger disincentives for work than a fully unconditional UBI).[22] The slight reductions in labour supply overall were mainly explained by the behaviour of specific groups: parents took more time out of the labour market to look after their children; and young people were more likely to stay on in education, to improve their skills. Need I point out that these are all things that the state currently subsidizes people to do, at considerable cost, because they are felt to be socially desirable? In short, as Michael Howard has put it: '[In the pilot schemes] people withdrew from the labour market, but the kind of labour market withdrawal you got was the kind you would welcome'.[23] In more recent trials of a full

---

22 See Widerquist, K. (2005). A failure to communicate: What (if anything) can we learn from the negative income tax experiments? *Journal of Socio-Economics* 34: 49–81, https://doi.org/10.1016/j.socec.2004.09.050

23 Quoted in Standing, G. (2017). *Basic Income: And How We Can Make It Happen* (London: Penguin, p. 163).

UBI in India and Namibia, overall economic activity actually went *up*, as more people were able to afford to access job markets, or began entrepreneurial activities on their own accounts.[24] I believe that under a UBI scheme, work would continue, and become better: innovation, worthwhile work, scholarship, and the arts would flourish, whilst degrading or miserable jobs would have to pay people more or treat them better. Hardly the end of civilization as we know it then.

If people persist with their intuition that UBI incentivizes people to do nothing, then the argument of last resort is the following: If you think it is stupid to give money to people even if they do nothing (UBI), then you ought to think it *really* stupid to give people money only on condition that they do nothing (the current means-tested benefits system). How much sense does that make?

§

There is one other great obstacle to acceptance of the UBI. People can't figure out whether it is a left-wing idea, or a right-wing one, so neither side takes it fully to its heart. At first it seems left-wing: making the welfare system more humane and less conditional, transferring money from those with most income to those with less, is the latest tool to further a long-standing socialist or social-democratic concern with inequality and social justice. The neoliberal big idea has failed. A big idea based on collective action must replace it, and the UBI is part of that idea.

But good UBI arguments have come from the right, too. Free-market economist Milton Friedman flirted with the idea, and the most serious Federal-level US policy initiative, the Family Assistance Plan (born about 1968, died about 1973) was proposed by a Republican president (Nixon) and largely killed off by the Democratic party.[25] The right-wing (or libertarian) argument is that UBI massively simplifies the state, and could facilitate it relinquishing a lot of its micro-control over our lives. For example, if a UBI is there providing a protective floor for everyone,

---

24 Basic Income Grant Coalition, Namibia. (2008). *Basic Income Grant Pilot Project Assessment Report*. Dowloaded from: http://www.bignam.org/Publications/BIG_Assessment_report_08a.pdf; Davala, S. et al. (2015). Basic Income: A Transformative Policy for India (London: Bloomsbury), https://doi.org/10.5040/9781472593061

25 The Family Assistance Plan was not a full UBI—it was something closer to a negative income tax—but it did represent a move in the UBI direction.

does the state also need to regulate minimum wages so closely? Couldn't people—protected from dire exploitation by the UBI—make their own minds up about what paid labour they wish to do under what conditions? Perhaps, going further down this line, the UBI plus control of law and order, is pretty much *all* the state needs to do, internally at any rate. We've given everyone enough to avoid starvation and be able to participate in economic life in a minimally sufficient way. After that, they are on their own: they can contract for the goods and services they choose in the market. This argument makes UBI the missing piece that completes, not replaces, the neoliberal vision.

In another essay, I have written about the difficulty of inter-disciplinarity.[26] Valuable integrative ideas can languish in the academic uncanny valley—not obviously owned by one discipline or another—and thus fail to have their potential recognized by anyone. Ideas that are quite good from two points of view, perversely, end up being championed by neither side, and thus have less immediate success than ideas that only appeal to one camp or the other. But what happens to the best of these ideas, in the end, is interesting: They go quite abruptly from all parties saying 'that makes no sense', to all parties saying 'well, everyone knows *that!*'. There's a similar adage in public policy: Important policy reforms are politically impossible, until just about the point where they are politically inevitable. We've seen plenty of examples of this in the slow and halting march of progress. Perhaps that is what will happen with UBI. We will look back and wonder what took us quite so long. Until then—and this is what scholars are uniquely placed to do—we have to keep the idea alive.

---

26 Waking up and going out to work in the uncanny valley, this volume.

PART THREE

11. The need for discipline

> ...far from characterizing [academic
> disciplines] by theoretical, disembodied
> abstraction, I view them as sites for the
> coordination and embodiment of skill.
>
> –Timothy Lenoir[1]

On and off over the years, I have sung in choirs of various types. I get by just fine, as long as I can stand in the middle of the basses somewhere, carried with the rumbling tide, but I am not a very good singer, or likely to ever become one. So why do I do it? Well, there are the obvious things: it's a bit social; it keeps me off the streets of an evening; and—not to be overlooked—it facilitates getting to know and love pieces of music in a deeper way than having the radio on will ever do. But none of these is the thing I enjoy most about choirs. The thing I enjoy most—and this is something I have not publically admitted before—is watching the choir-master (or choir-mistress) do their work.

If you have been to a choral rehearsal, you will know what I am talking about. The choir-master will be sight-reading the piano reduction of the orchestral score, switching what they are playing to bring out one or the other of the choral lines, all the while listening to four different sections singing, spotting the kinds of difficulties they are having, and remembering tips that need to be given afterwards. The choir finishes a movement. I am just pleased to have got to the bottom of the page at about the same time as everyone else. The choirmaster, however, pipes up: "I was wondering in bar 73 if the diminuendo should be in the sopranos only; the $c^{\sharp}$ in the alto line heralds the key change we are going towards, and we need to hear it". And here I am thinking: *how were you wondering anything in bar 73?* You were sight-reading three lines of music and singing a fourth, whilst simultaneously attending to the

---

1   Lenoir, T. (1997). *Instituting Science: The Cultural Production of Scientific Disciplines* (Stanford: Stanford University Press, p. 2).

     https://doi.org/10.11647/OBP.0155.11

timing and pitch and volume and pronunciation and breathing of about 40 different people. It takes me about 3 minutes just to work out where bar 73 is. Yet you did all you were doing so effortlessly that you still had capacity spare. It's a mental and physical dexterity with the components of the music that I can appreciate but never reproduce. "Basses, you seemed uncertain on that *b* natural entry; you are getting swayed into pitching it flat because of the *e* flat in the tenors. Listen the end of the sopranos' tune the bar before, and think, 'happy *birth*day to you'". Will do, boss. Which ones are the sopranos again?

Yes, there's something fascinating—moving even—about seeing people exercising real skill. You might think that observing the highly skilled would be alienating or aversive for those of lesser skill. In fact, the opposite seems to be the case. We flock to hear virtuoso musicians and watch master chefs on television; we seek out dry-stone walling competitions, sheep-dog trials, and demonstrations of glass-blowing. I remember once Melissa and I were having a complex-shaped roof covered in lead on the back of our terraced house. The roofer gadgie duly turned up with his bag of lead hammers and his rolls of lead. I said to our next-door neighbour that I hoped that the builders were not disturbing his peace. On the contrary, the neighbour said—I cannot confirm but would like to imagine a flutter seizing his breast—it's such a privilege to be able to see that man *work*.

It's a privilege to see a skilled person work because, I would contend, it connects us to something deep about being human. We are the species that is good at getting good at doing stuff that is hard to do. We do this rather eccentrically and ecumenically: sometimes in domains with a utilitarian payoff, sometimes not. Some individuals take skill further than others in any particular domain, and this capacity for individual specialization is itself interesting and consequential. But skill acquisition is not the preserve of a few geniuses: it's just what human beings do. It is because it is so pervasive that we only notice the extreme cases. By the time chimpanzees are a few years old, they have pretty much reached peak productivity, but humans—their close evolutionary cousins— live their lives by expressing extraordinary skills that can take twenty, thirty, forty years to develop and refine.[2] I have an obscure and possibly

---

2   Kaplan, H., K. Hill, J. Lancaster and A. M. Hurtado. (2000). A theory of human life history evolution: Diet, intelligence, and longevity. *Evolutionary Anthropology* 9: 156-85, https://doi.org/10.1002/1520-6505(2000)9:4<156::aid-evan5>3.3.co;2-z

sentimental sense that it is in the exercise of these embodied skills that humans reach their fullest sense of personhood.

It upsets me that our living species gets the name *Homo sapiens*, the human that knows. Watching my choirmaster, or the roofer gadgie, what strikes me is not *that they know* so much as *that they can do*. What they have is not knowledge in some purely propositional sense, something that could be stored on a USB stick. If it is knowledge, it is procedural knowledge, instantiated in and distributed across the whole of the body, and realised in patterns of movement (striking the keys, bending the lead, tensing the diaphragm, modulating the larynx). Indeed, it's often knowledge that cannot be expressed explicitly or imparted verbally, so highly routinized and embodied has it become. We lost out in the naming game to our extinct relative, *Homo habilis*. The skillful human. That's what I would like us to be called. I would take *being able to do* over just *knowing* any day of the week.

What name should we give the virtue that we recognize in highly skilled people? I would, for the sake of today, like to call it *discipline*. Discipline, in the everyday sense of self-control, is what is needed to drive oneself through the 40,000 hours of practice and training that high skill requires. But the word *discipline* has broader resonances: it links back to the Roman deity *Disciplina*, with her virtues of skill, self-improvement, economy of action, dedication to the guild, and simplicity of life. What could be more attractive, then, than *discipline*?

§

I began with a paean to discipline in order to wrong-foot you. If you are anything like me, you probably have a well-developed sense that in science, *disciplines*, entailing as they seem to *disciplinary boundaries*, are a bad thing. I have spent my whole career railing against them. Indeed, it's something of an identity marker amongst my people to deplore the balkanization of the study of human behaviour across so many discrete disciplines; to blame disciplinary divisions for our failures to progress; and to claim to be trans-disciplinary or post-disciplinary, in our orientations. We don't tend to say much about how the landscape of our post-disciplinary utopia ought to be organized for practical purposes; only that, perhaps rather suspiciously, a lot more status and resources would be accorded to people who…well, to people who are like ourselves, really.

Often in academia, we clarify our reservations about some idea by stating that idea in its most stark or simple form: the famous *straw man* strategy. But a very useful complement to the straw man strategy is the *steel man* strategy: try to characterize the idea you oppose in its best, most sophisticated possible form.[3] If you can defeat even the steel man version of an idea, then it really is a bad one. More likely, you will discover unappreciated virtues in an idea that you previously thought of as wholly bad, and adopt a more nuanced position. I feel like this about disciplines. I have read (and written) so much argument against the segmentation of the academy into discrete disciplines. Yet academic disciplines got invented, have been perpetuated by a lot of very clever people, and largely continue to exist; indeed, science has been doing conspicuously well since about the time disciplinary structures became established. All of which leads me, in the spirit of the steel man strategy, to ask: what is there that is *good* about disciplinary structures?

With this question developing in my mind, it was naturally with interest that I listened, in a restaurant in Helsinki as it happens, to a friend telling me about how a certain university was reimagining the structure of their curriculum. Instead of organizing courses of study around discrete disciplines, students would study a portfolio of modules whose subject matter was defined by phenomena or problems, like migration, climate change, or violence. The problems chosen were exemplary, in that their solutions could not be generated by any discipline acting alone. Within each problem, students would learn how economists thought about it, how sociologists thought about it, how biologists thought about it, and so forth. This would give them the ability to compare, contrast and syncretise different perspectives without being artificially shunted into the confines of any one of them. Great, I thought, inanely: exactly my kind of thing. 'How does it work in practice?' 'Terrible', she replied. Students could produce generalized and often stereotyped comparisons of different disciplinary approaches, but without enough depth or detail to actually implement (or improve)

---

3   I am grateful to Brett Beheim for introducing me to the idea of the steel man strategy.

any of them. They were left with the abiding impression that what you believe about the world is really just a matter of what identity you choose to adopt, rather than the consequence of systematic epistemic work using justified standards. In short, they came out of their studies *not knowing how to do anything*.

This anecdote reminds us that the origin of the term *discipline* in the academic context is a pedagogical one: the set of training you need in order to be a competent and useful practitioner in a domain. And this training is not reducible to the acquisition of factual statements; not even reducible to the acquisition of factual statements plus frameworks for interpreting and explaining them. Perhaps more than either of these things, the concept describes a set of core physical skills. In this sense, the 'discipline' of the mathematician or ethologist has more in common with the 'discipline' of the roofer, the stonemason, or the choir-master than one might at first imagine. When we think about practical skills, inter-disciplinarity does not seem like a particular virtue. It might not be bad, but it is a lesser virtue than excellence within the relevant domain. Who would you hire: the inter-disciplinary welder — 'I can weld a bit, and I can critically compare welding to carpentry!' — or the welder who welds well?

§

It might be useful to separate analytically two components of scientific disciplinarity: the *declarative* and the *practical*. On the declarative side, academic disciplines sustain particular structures of explicit belief and understanding, and they do so in part through political and ideological operations. (I am not saying that science is mere ideology or mere politics — on the contrary, its content is in the end constrained by nature. It is nonetheless a social process that proceeds year to year through ordinary human manoeuvring.) Disciplines exert power, and sustain ideologies, through control of what gets published (and in what form); through control of funding panels; and through control of academic hires and curricula. By such means, disciplines can define what types of question can be asked, in what way, and what constitutes an acceptable answer. They provide handy non-reasoned authority and legitimation for particular decisions and inferences. I remember one conversation with a colleague about why she was interpreting a particular behavioural phenomenon as evidence for a particular cognitive mechanism in the children she studied. Her reply was simply that, in developmental psychology (her discipline), that's how researchers interpret it. Whilst her answer was pragmatically realistic, it was epistemologically unsatisfactory. I was tempted to recall George Berkeley's jibe against scholastic philosophers: 'when a Schoolman tells me 'Aristotle hath said it', all I conceive he means by it is to dispose me to embrace his opinion with the deference and submission which custom has annexed to that name'.[4] Judging this a bit heavy for a Tuesday lunchtime, I held my tongue.

It's the declarative aspects of disciplines, particularly the way they trammel and police researchers' explicit cognition, that are the easiest to use in an indictment against them. They normalize assumptions that should be exposed, provoke cognitive conformity, and delimit possible moves and juxtapositions. Thus, they blind us to aspects of the phenomena, or theoretical resources, that might hold the key to progress. I'll come back to this argument, which is the one I have habitually relied on in advocating post-disciplinarity. For now, let us concentrate on the fact that this argument gives no recognition to the practical components of disciplines.

The practical components of disciplines are the physical skills they serve to inculcate, transmit and refine. When you hire a mathematician, you want them to be able to do matrix algebra, and when you hire a molecular biologist, you want them to be able to pipette. You can expect that they will be able to do so skilfully in virtue of the disciplinary training that they have received. Disciplines in this sense are guilds of artisans. It seems obvious that to drive levels of skill higher and higher, there will need to be specialization, deep apprenticeship, and assortment of artisans with the others of their guild. The wood turners may get on well with the potters, but they will naturally want to spend a lot of time with other wood-turners to mutually improve their skill, and to pass it on. Viewed from the practical rather than the declarative perspective, then, the existence of separate disciplines seems much less pernicious and much more natural. And indeed, historically, it is

---

4   Berkeley's jibe is from the introduction to his 1710 Principles of Human Knowledge (various editions).

the development of particular laboratory or field practices, as much as declarative theoretical commitments, that gives rise to new disciplines.[5]

You might concede the need, on practical grounds, for disciplinary specialities such as electrophysiology or molecular biology. You might however still cling to the view that the broad domain covered by psychology, sociology, anthropology and economics should simply be one open field, since these disciplines all look at the same thing, namely human behaviour. That's only right to an extent. After all, electrophysiology, fMRI and EEG all look at the same thing—neural activity—but nonetheless require discrete skills. Capturing and analyzing large amounts of data on monetised transactions; getting to know and understand the lives of a particular small social group; designing and interpreting large-scale attitudinal surveys; making experiments to isolate the causal structure of particular cognitive processes: these are different operations that each require deep reservoirs of skill to do well. The lack of expensive equipment and white coats should not blind us to the complex and distinctive skills required in each case. And I know from experience how easy it is, as a disciplinary novice, to do these things so badly that the results are basically useless.

§

Another musical anecdote: Melissa and I play baroque recorder duets. She's the more skilful player. I am known for proclaiming, as we embark on a new piece: I think this movement should have a *misterioso* quality. Her translation: what you mean is that it's too difficult for you to play properly. Your *misterioso* is basically made up of slowing down on the hard bits, getting some of the notes wrong, and a distracting smokescreen of emphatic head movements. When I practice a bit, guess what, I discover that it sounds even better if, rather than *misterioso*, it is just played well.

---

5   For this view of the history of science, see Lenoir, T. (1997). *Instituting Science: The Cultural Production of Scientific Disciplines* (Stanford: Stanford University Press). The shift from seeing scientific disciplines as sets of declarative knowledge to seeing them as sets of skills is echoed in a recent paper suggesting the same move in the study of human language. What we acquire when we learn a language should not be seen primarily as a set of abstract declarative principles, so much as the practical ability to understand and produce speech in real situations: Chater, N. and M. H. Christiansen. (2018). Language acquisition as skill learning. Current Opinion in *Behavioral Sciences* 21: 205-8, https://doi.org/10.1016/j.cobeha.2018.04.001

In like fashion, I have been involved in several research projects that were self-consciously trans-disciplinary, usually mixing some kind of evolutionary biology theory with some kind of social science data. As we accumulated rejection slips from the best journals in the field, we consoled each other with the insight that journal editors and reviewers are cognitively trapped in disciplinary silos, and prejudiced against any kind of attempt to transcend them. Our indignation affirmed our sense that we were right and what we were doing was important. Indeed, at times, it seemed like the more the disciplinary specialists rubbished it, the more vindicated we felt in our worldview. I shouldn't need to tell you that this is an extremely disturbing epistemic direction to be headed in. Looking back over this history now with the benefit of a bit more experience, I can see a plainer truth: the studies we were doing often contained interesting germs of ideas, but were *just not very well done*. Our trans-disciplinarity was, not always but all too often, the *misterioso* of science: a pretext for sketchy methods, careless design, hasty data analysis, inferential over-extension, and lack of theoretical precision.

Pat Bateson, nearly fifty years ago now, distinguished between roundheads and cavaliers in the study of behaviour. The roundheads are methodologically impeccable in every respect, but, in his words, unwilling to flirt, let alone dance, with ideas. The cavaliers exhibit great dash and intellectual exuberance, but this does not come without cost: they 'also are notoriously unsound and constantly confuse inference with evidence'.[6] It would be too much of a stretch to equate roundheads with disciplinary specialists and cavaliers with trans-discipliners. There can after all be people who are cavalier within the bounds of one discipline, and trans-discipliners who develop beautiful and rigorous methods. There is no doubt in my mind, though, that across the study of behaviour, greater trans-disciplinarity comes with a more cavalier attitude to methods on average. Which of course links us back to the practical components of disciplines.

I can think of many examples from my career where the growth of a trans-disciplinary research area produces high-impact publications that could have been better with a little more discipline. There are the famous behavioural-economic experiments demonstrating that people have a pro-social concern for others' financial outcomes as well as their

---

6   Bateson, P. (1970). What is learning? *New Scientist*, 25 June: 621–3, p. 621.

own. These experiments consist of artificial financial dilemmas, in which many participants can choose (or not) to pay into a group fund that will benefit everyone at net cost to them. I firmly believe, by the way, that humans do have other motives than just maximizing personal gain, including prosocial motives. It's just that these particular experiments could have been designed better, and provided stronger grounds for their conclusions. Specifically, the original experiments did not include control conditions to separate the motivation that others should benefit from, for example, failure to truly understand the rules of the game, or a dislike of using the extremes of a scale. Once you include control conditions to rule these alternative influences out, the evidence that participants' behaviour reflects concern for others' outcomes becomes much less convincing.[7] The original studies—published in the highest-profile inter-disciplinary journals—were done by brilliant economists who lacked deep discipline in experimental psychology. Experimental psychology has many faults, but one of its virtues is real skill in designing experiments and, in particular, the almost ubiquitous need for multiple control conditions in order to make inferences about the meaning of an experimental effect.

To take another example, in primate neuroscience and primate cognition, it's common to find people taking 1000 or even 5000 trials to train their animals on a simple discrimination (say, between two colours) where one option is rewarded and the other not. Even after thousands of trials, performance is not always very good. There is even a view out there that smart animals like chimpanzees do not readily acquire arbitrary discriminations in the way that rats or pigeons do.[8] The fact that these monkeys and apes take so many hundreds of trials, though, is an artefact of the way they are trained. Intuitively, it seems like the way you would make an animal learn an association between a stimulus and a reward is by pairing the two as often as possible; so in these training paradigms there may be hundreds of pairings of stimulus

---

7   Burton-Chellew, M. N. and S A. West. (2013). Prosocial preferences do not explain human cooperation in public-goods games. *Proceedings of the National Academy of Sciences* 110: 216–21, https://doi.org/10.1073/pnas.1210960110

8   See Hanus, D. and J. Call. (2011). Chimpanzee problem-solving: Contrasting the use of causal and arbitrary cues. *Animal Cognition* 14: 871–8, https://doi.org/10.1007/s10071-011-0421-6; and Bateson, M. and D. Nettle. (2015). Development of a cognitive bias methodology for measuring low mood in chimpanzees. *PeerJ* 3: e998 for some discussion, https://doi.org/10.7287/peerj.preprints.888

and reward, with very short intervals between them, in a single session. It turns out that intuition is spectacularly wrong in this particular case: the way you make an animal learn an association between a stimulus and a reward is to make the stimulus, and hence the reward, very rare (i.e. long gaps between trials). Once you do this, the number of pairings required comes down by orders of magnitude.[9]

Old-school rat and pigeon animal learning theorists knew this very well, and they could train an arbitrary association perfectly in a dozen or two trials. They also knew how best to navigate the way to learning a full discrimination: first training the association between the positive stimulus and reward, then introducing the unrewarded alternative, then building up to choices. Just sticking two colours in front of an animal again and again in quick succession is only going to teach them something by a near-endless war of attrition; the unrewarded stimulus ends up temporally proximal to the rewarded one, and the required informative contingencies in experience are lacking. So the interesting question is why the skill held by animal learning theorists has not found its way into the communities studying primate neuroscience and primate cognition. Well, by and large those researchers don't have deep discipline in animal learning theory. Instead, they are coming in from anthropology, or if it is from psychology, it is cognitive psychology. One of the sad and pointless things about the 'cognitive revolution' in psychology, in which behaviourism was allegedly 'overthrown', is that a lot of really useful skill in how to make animals learn, as well as how to design beautiful experiments, was lost in a kind of year-zero mentality. We need to build on the practical skill of behaviourist psychology, not throw it out with the cognitive bathwater.

To take a final example, in the last five years, I have begun to work in telomere biology, not really because I know much about telomeres, but because of the possibility they offer us to provide an integrative marker of the insults and damages inflicted by the world over the course of an individual's life. We did four successive experiments where we showed

---

9   This phenomenon becomes comprehensible once you appreciate that what you want to maximize is not the number of times that a stimulus has been paired with a reward, but the information an event carries about a reward. Rare events carry more information. See Ward R. D., C. R. Gallistel and P. D. Balsam. (2013). It's the information!. *Behavioural Processes* 95: 3–7, https://doi.org/10.1016/j.beproc.2013.01.005

that telomeres shorten very rapidly in the nestling starling, then a fifth experiment where they did not—they appeared to get longer, in fact. Had we discovered an extraordinary exception that would form the basis of a *Nature* paper? Reverse the ageing process! A group of Northumberland starlings holds the key to immortality!

I showed the data to a colleague I admire who is more skilled than me in the actual lab work. She didn't need to see the results of my multi-level model; she didn't even need to see the statistics on the technical replicates. She certainly didn't need to hear my elaborate theoretical interpretation. 'Oh', she said, 'your reaction hasn't worked. Look, those numbers are too high. Your primer concentration must be wrong'. And that was it. She was like a choir-master, knowing in her bones where and why the basses had been misled. The basses have not stumbled on an interesting new direction for Western music; they are merely singing badly. The methods sections of the published papers on measurement of telomere length by quantitative PCR say all kinds of useful things, of course, but what they don't say is: if these numbers aren't quite a lot smaller than those other numbers, you've probably just done it wrong. And in the emerging trans-disciplinary field of telomere epidemiology, a number of the most exciting 'findings'—such as the idea that longer telomeres shorten faster, or that over short periods, the telomeres of about 50% of people get longer, probably reflect measurement error as much as anything else.[10] If we had all had a bit more discipline prior to publication, perhaps science could have proceeded more efficiently.

§

The conclusion so far seems to be that we should want to retain all the practical virtues of disciplines, namely having people with high levels of specific technical skill, but abolish the declarative distinctions between them. I don't know if this is possible; as Timothy Lenoir has argued, the connections between the practical and the declarative components

---

10   See Steenstrup, T. et al. (2013). The telomere lengthening conundrum — artifact or biology? *Nucleic Acids Research* 41: e131, https://doi.org/10.1093/nar/gkt370; and Verhulst, S. et al. (2013). Do leukocyte telomere length dynamics depend on baseline telomere length? An analysis that corrects for "regression to the mean". *European Journal of Epidemiology* 28: 859–66, https://doi.org/10.1007/s10654-013-9845-4

of scientific programmes are intimate.[11] Indeed, I have often wondered to what extent the explicit propositional commitments of particular disciplines arise spontaneously from the types of practical activities that their work involves. If you spend all day in dealing with prices and purchases between anonymous actors in a fungible currency, maybe you start to think like an economist. Maybe you would do so even if you had not been indoctrinated in micro-economic theory; even if that theory did not already exist. If by contrast you spend all day in open-ended non-monetised interaction with a small group of people, maybe you either need to lie down in a very quiet room, or you start to think like a social anthropologist, or both. On this view, the declarative differences between disciplines would not be (just) historically contingent ideologies sustained by the dynamics of power and influence, but the inevitable *déformations professionelles* arising from individuals habitually working at different practical activities. Hence, there is no clear dividing line where the concern that the practical activities be done skilfully ends, and ideological boundary-maintenance begins.

So it's going to be hard work to overthrow the declarative balkanization of science without any loss of practical skill. And actually, allowing the steel man to be even steelier, there exists an interesting literature arguing that some declarative balkanization of science is a good thing. The argument (which I am interpreting in my own, slightly *misterioso*, manner) goes something like this: the progress of knowledge relies on variation, for exactly the same reason that adaptation by natural selection does. We need to be trying out a lot of different ways of thinking. It is difficult for the same individuals to entertain more than one way of thinking simultaneously. Thus, the simultaneous existence of multiple *groups* of individuals, each thinking about human behaviour in a different way, is actually a strength. The internal coherence of the groups is not undermined by constant blending, and at the meta-level, competition between the groups for society's attention and support is an invisible hand propelling humanity towards a higher level of overall understanding.[12]

11   Lenoir, T. (1997). *Instituting Science: The Cultural Production of Scientific Disciplines* (Stanford: Stanford University Press).

12   For versions of this argument, see Stichweh, R. (1992). The sociology of scientific disciplines: On the genesis and stability of the disciplinary structure of modern science. *Science in Context* 5: 3–15, https://doi.org/10.1017/s0269889700001071; and

This kind of argument comes up in domains other than science. The economic dynamism of early modern Europe (in contrast to China, say) has been attributed to the existence of many independent city states, in competition with one another, where different things could be tried out: the good things could spread, and the bad things didn't drag the whole continent down. Political devolution in the United Kingdom has had the virtue that new policy ideas are tried out independently in Scotland or Wales, with adoption more broadly dependent on the results of those innovations, which are effectively natural experiments. In science, this kind of progress-through-variation depends on the disciplines being somewhat informationally isolated (or all the variation would be rapidly washed out), but nonetheless a little bit leaky. For all of the United Kingdom to benefit, the idea that works well in Scotland does have to find its way to England in the end. Similarly, the ideas of behavioural ecology need to be able get to sociology (and vice versa), but without sociology just becoming behavioural ecology.

What we seek then, if this line of reasoning is correct, is a goldilocks level of trans-disciplinary integration: not too little, not too much, but just right. The optimal level would be something that could be modelled. And much as people like me moan on that the current level is too little, the solid evidence that this is the case (for example, that science progresses faster in periods or areas where disciplinary integration is greater) is currently lacking. Indeed, bibliometric studies have generally concluded that inter-disciplinary leakage is already quite substantial and ubiquitous under the status quo.[13]

§

Having spent a few days with the steel man argument for disciplines, has my commitment to greater trans-disciplinary integration changed? In one sense, no: there is still so much that has not yet been done, but

---

Jacobs, J. A. and S. Frickel. (2009). Interdisciplinarity: A critical assessment. Annual *Review of Sociology* 35: 43–65, https://doi.org/10.1146/annurev-soc-070308-115954. For a recent formal model of why epistemic diversity leads to better discovery of knowledge about the world, see Devezer, B. et al. (2018). Discovery of truth is not implied by reproducibility but facilitated by innovation and epistemic diversity in a model-centric framework. *ArXiv* 1803.10118, arXiv:1803.10118v2

13 See Jacobs, J. A. and S. Frickel. (2009). Interdisciplinarity: A critical assessment. Annual *Review of Sociology* 35: 43-65, https://doi.org/10.1146/annurev-soc-070308-115954 for review.

could be if we could draw our resources and assets together. But I have perhaps tempered my views. Not everyone should be bludgeoned into inter-disciplinarity all the time; and when we undertake it, we should undertake it constructively. The thinking styles of other disciplines, arbitrary though they can appear, have arisen for particular reasons, often to do with the reality of the practical activities in that area, and these reasons need to be understood. It's perhaps inevitable that there will always be cavaliers and roundheads, and that the first to rush into a new area will be more cavalier than those who come along later to tidy up. Nonetheless, we can try not to be *too* cavalier. The disciplinary specialists have good practical reasons for apparently niggling concerns.

Modish desires for inter-disciplinarity should not trump our commitment to training people to high levels of specialist skill. Some of the key skills that we need to train—statistical modelling, for example—are already trans-disciplinary anyway, but others are specific to particular kinds of data. The glitter of inter-disciplinary declarative statements can, for certain types of personality, be more attractive than the grit of intra-disciplinary practical skill, but at least some of the latter is indispensable if one wants to get anywhere.

Rather than viewing other disciplines as competitors to be rhetorically trashed, or ailing companies susceptible to a hostile merger, we should try to understand what skills those disciplines embody, and see if we can get access to those skills to raise the level and refinement of our research. This is something that can be done by training, and/or by collaboration. The best way of understanding what a discipline has to offer is not just to read its theoretical end-products, but to attend to the detail of its practical methods and processes. Ideally, it's good to gather and handle the kinds of primary data that a particular discipline deals with; this might illuminate why its practitioners have the particular concerns and notions that they do. If you want to build deep cross-discipline links, my hunch is that you will get further by spending time working in your collaborator's lab (or field site, or archive) than you will exchanging formal talks about each other's high-level belief systems. The latter activity can go two ways: it can become polarising, entrenching the perception of difference and incommensurability; or it can lead to a superficial patina of consilience without the two disciplines really growing into one another.[14] Trying to do primary work together, on practical problems, creates living connective tissue that is properly vascularized from both directions.[15]

What should the academic landscape look like at the macroscopic scale? What we seek is what Kevin Zollman calls 'transient epistemic diversity'.[16] Students of social life need to be able to pursue the phenomena of social life without having to wait for the physicists to reconcile the strong and the weak nuclear forces first. They need to be allowed to build up and stabilize the best possible skills for doing so. On the other hand, the modules so formed should never become completely hermetic. If our social theories are incompatible with the laws of physics or biology, we can't just shrug and say, 'not our problem' indefinitely. How can we realize this paradoxical, unity-in-plurality, world?

Something we can draw on here is the idea of a small-world network.[17] Small-world networks manage to have two interesting properties simultaneously. The first is a high degree of clustering or cliquishness: most interactions are local, and most of a focal individual's interaction partners are also interaction partners of one another. Thus, small-world network architectures would allow for domain-specific transmission of skill—for guilds, if you will. The second property is a surprisingly short path length connecting any two nodes (the famous six degrees of separation/Kevin Bacon/Paul Erdős). Thus, if you put information into a small-world network at any point, it's really not long before it shows up everywhere (if it's spreadable information that is). Small-world-ness turns out to crop up in evolved biological systems, like nervous systems, presumably because these have similar requirements as knowledge does: specific sub-parts need to deal with specific problems through local specialization, but the whole thing also has to function somewhat coherently at the macroscopic scale.

---

14 The importation of 'life-history theory' from evolutionary biology into psychology is perhaps an example. See Is it explanation yet?, this volume.

14   The importation of 'life-history theory' from evolutionary biology into psychology is perhaps an example. See Is it explanation yet?, this volume.

15   An argument set out by Watts, D. (2017). Should social science be more solution-oriented? *Nature Human Behavior* 1: 0015, https://doi.org/10.1038/s41562-016-0015

16   Zollman, K. J. (2010). The epistemic benefit of transient diversity. *Erkenntnis* 72: 17-35, https://doi.org/10.1007/s10670-009-9194-6

17   Watts, D. J. and S. H. Strogatz. (1998). Collective dynamics of 'small-world' networks. *Nature* 393: 440-2, https://doi.org/10.1038/30918

How do you make a small-world network? It's simple. You have a grid of nodes (researchers in this instance). You set up a lot of short connections between neighbours, and a smaller number of long-distance ones to a random point elsewhere. From a small-world network point of view, debates about inter-disciplinarity are simply debates about whether a parameter p of the network architecture (the proportion of connections that are long-distance rather than local) is currently too low (too much hermeticism) or too high (insufficient epistemic specialization).

We would make science a small-world network if we trained all people deeply in a disciplinary tradition, but encouraged many of them to spend at least one year, at some point in their careers, in a completely different discipline, getting research training and actually doing some stuff. I don't think this teleported year should be restricted to students starting out; full professors could benefit from it too. The disciplinary combinations could be as bizarre as you like: the way you get small-world-ness in a network is exactly by any long-distance link whatever being possible with a certain probability. We need to provide mechanisms, within academic careers, for these long-distance links to be made, and the extra skills acquired, but not at a cost to the depth of formation in whatever people initially do.

Another mechanism we could employ to maintain small-world-ness is complementary peer review. I don't mean your reviewers should compliment you on how great your paper is, though that would indeed be refreshing. I mean that every paper, during its development, should receive two complementary inputs. The first is a detailed assessment of the methods. You are really only going to get this from people who have practical skills in the right domain. These reviewers are going to be responsible for driving the methodological rigour, in a purely local sense, higher. The second input is from someone in a distant discipline. The point of this review is to say: did you know people have already thought about this kind of problem in this other literature, and they tend to think about it in this way; or, we've actually got a method over here for modelling that situation—why don't you incorporate it; or just, why do you assume *that*? The local review would hold the work to account against the practices of the discipline; but the distant review would also hold the whole paradigm to account against the rest of knowledge. Both of these are important functions. The distant review might also have the by-product benefit of reigning in bad writing and jargon, since authors would have to describe their work comprehensibly enough that the distant reviewer could understand what they were talking about. Both of these reviews could be published alongside the paper, essentially tying in both the disciplinary experts and the neighbouring fields in the common cause of trying to understand the world better.

At present, you tend to get only the local review in specialist journals, which is why these journals are filled with papers that are locally adequate but conceptually derivative, and sometimes only the distant review in inter-disciplinary journals like *Nature* and *Science*, which is why these journals contain more than their fair share of methodologically unsound research. I personally think we should all publish in some big online archive that spans all subject areas, and features both local and distant open peer-review.[18] And I think, more generally, the onus of peer-review should be moved away from *rejecting* pieces work, and more to collaborating constructively on their development, starting where possible before the data are actually collected. If reviewers could make the conceptual leap from anonymous controllers of access to a scarce and zero-sum resource, to critical co-producers of shared information about the world, surely we would all be better off.

A final point about small-world networks. Although they have clusters, they have no abrupt cluster boundaries. One dense region gives way continuously to another. And this was of course true of early modern Europe too. It was only relatively recently that passports, borders, and border guards existed.[19] Prior to that, there were certainly centres of influence and innovation, each with the ability to generate its own laws and norms, but their influence shaded off with distance like gravitational fields. Many people lived in marches or borderlands with access to more than one political centre. Political centres influenced one another and were involved in higher-level federations. Travel, though physically hard, was not administratively policed. Perhaps there are ways of maintaining discipline, but without boundaries.

---

18   I think this should be controlled by a self-organizing not-for-profit collective, much as the programming language R is.

19   Carr, M. (2013). Beyond the border. *History Today,* January. Downloaded from: http://www.historytoday.com/matt-carr/beyond-border

12. Waking up and going out to work in the uncanny valley

> ...giving up a compact disciplinary
> identity can be very risky.
>
> –Rudolf Stichweh[1]

Film folklore has it that, in François Truffaut's film *Tirez sur le pianiste* (1960), Charles Aznavour's central character never actually occupies the centre of the frame. Whether or not this is quite true, he certainly spends a lot of time round the edges, down the bottom, or out of shot entirely. It's an apt visual mirror: there's a gap between his great artistic aspirations and the reality of his achievements. He has an air that the attention has always moved somewhere slightly different from wherever he is.

My academic life is likewise permeated with a constant sense of slight marginality; of failure to ever get myself quite to the middle of the frame. This may be just my personal mixture of insecurity and self-importance. Or perhaps *all* academics out there feel that they are more peripheral than everyone else. For example, the Psychology degree programmes I have worked on are periodically audited by the relevant professional body. Part of the body's concern is with how much of the teaching is done by 'real' psychologists. I've always felt vulnerable here: I have slightly less than half a degree in Psychology, my PhD is in Anthropology, and I don't usually publish in journals with 'Psychology' in the title. However, asking around, it seems like pretty much all of my colleagues also feel that, for one reason or another, they are not 'real' psychologists either. And what is more, they seem to see *me* as the real psychologist!

1   Stichweh, R. (1992). The sociology of scientific disciplines: on the genesis and stability of the disciplinary structure of modern science. *Science in Context* 5: 3–15, p. 13, https://doi.org/10.1017/s0269889700001071

https://doi.org/10.11647/OBP.0155.12

Even if everyone feels a bit off-centre, though, I feel a long way off-centre sometimes. I have ended up, by random stumbling around as much as by judgement, living my whole life outside the comforting shelter of any single disciplinary or sub-disciplinary encampment. Two consequences of this strike me as non-obvious enough to dwell on. The first is the following: I have an easier time talking to colleagues about the components of my work that are far from their concerns, than the components that are near to their concerns. The second consequence is that parts of my work are consistently misinterpreted or misremembered as saying something that they really don't quite say. I offer these observations not (I hope) as mere whinges, but as reflections on something interesting about human cognition, about how it tries to impose categorical order on a shifting and continuous landscape of information.

§

Back in the 1970s, the Japanese engineer Masahiro Mori introduced the concept of the uncanny valley. The uncanny valley originally described a reliable phenomenon that occurs when robots are made more human-like in appearance and behaviour. As robots move from very un-human-like to a bit more human-like, people's psychological response to them becomes more positive and empathetic; and when the robots are completely indistinguishable from humans, we respond to them as humans. But there's a dodgy bit in between, in the place where the robots are getting really quite like humans, but occasionally leak cues that betray their artificiality. And in this gap—the uncanny valley—people don't like the robots at all. They like proper people, and they like good old-fashioned droids with crazy LED eyes and wires hanging out. They really do not like the thing in the middle, 'the thing that should not be'.[2]

---

2   There's a large literature on the uncanny valley phenomenon. I draw here in particular on: Saygin, A. P. et al. (2012). The thing that should not be: Predictive coding and the uncanny valley in perceiving human and humanoid robot actions. *Social Cognitive and Affective Neuroscience* 7: 413–22, https://doi.org/10.1093/scan/nsr025; and Ferrey, A. E., T. J. Burleigh and M. J. Fenske. (2015). Stimulus-category competition, inhibition, and affective devaluation: A novel account of the uncanny valley. *Frontiers in Psychology* 6: 1–15, https://doi.org/10.3389/fpsyg.2015.00249

Early accounts of the uncanny valley related it to our particular conception of 'the human', and our aversion to this key boundary being infiltrated or violated. But it turns out that the uncanny valley is a much more general phenomenon; you get one as you move continuously across many conceptual boundaries, not just the human/non-human one. So the explanation of the uncanny valley phenomenon needs to be rooted in more general ideas about how brains work.[3]

Brains are prediction machines. You can't do perception or cognition purely inductively, allowing the information in the incoming sense data to impress knowledge about the world onto a flat blank canvas. You can't do this because there are too many gaps in the immediate data: objects that are partly occluded by other objects; patterns of luminance that could reflect either a change of colour or variation in illumination; retinal images that could represent various combinations of size, shape and distance; ellipses and ambiguities in people's utterances. So brains need to create high-level 'models' of what is out there. These models employ categories that are at least to some extent discrete. Perception and cognition are as much as a case of your internal models projecting downwards to funnel the sensory input into some kind of structured form, as they are of the incoming information driving upwards to determine what you believe. That's why people are famously susceptible to all kinds of perceptual and cognitive illusions: their internal models can be tricked into firing erroneously in various ways.

The interaction of bottom-up sensory data and top-down internal model is a delicate one. Over the long run, internal psychological models are built up from experience and continuously modified by it, so it is the incoming data that determines the model in the end. (Or at least, the incoming data in interaction with inbuilt priors, such as you can't have two objects in any one place at a same time, something can't be both plant and animal, and so on.) Over the short run, though, the internal model provides a lot more weight than any momentary piece of experience. A one-off anomalous scene or object is therefore apt to be reinterpreted as something else, something more compatible with existing model schemas. The meeting up of incoming sensory data and

---

3   The particular view of how brains work described here comes from Clark, A. (2013). Whatever next? Predictive brains, situated agents, and the future of cognitive science. *Behavioral and Brain Sciences* 36: 181–204, https://doi.org/10.1017/s0140525x12000477. I am grateful to Rob Barton for introducing me to this paper.

internal model schema does not happen at any one stage in the neural hierarchy. Rather, there are many inter-connected processing levels in the brain. Each level passes down to the level below a prediction about what the world is currently like, and hence what data it should be receiving. The level below passes up one of two things: nothing, if the prediction is met and the world is as the model suggests, or an error signal, which effectively says 'No, it can't be that. What I'm getting deviates from that expectation in this particular way'. These prediction error signals do two things. Immediately, they cause the higher-level circuit to select another hypothesis about the world ('maybe that thing's not so close, it's just big; try this prediction'), and distally, they cause the model weights to be slightly adjusted so that next time, the circuit won't make the same mistake given the same cues.

How can we relate all this to the uncanny valley? Well, all is going well for your brain when it can choose an interpretation of the world that produces no residual error signal at all. What's this? Is it a goat? No, big error signal. Is it a person? Error signal equals zero. My work here is done. But there are some things that are uncanny, and this means, precisely, that they give you a big error signal whichever way you interpret them. Consider the faun. Is it a goat? No, big error signal. Is it a person? No, now I've got a different but equally large error signal. Damn. This thing is just…yucky on my brain.

Roughly speaking, people dis-prefer stimuli whose error signal can't be got down to a reasonable level, stimuli that defy sensible resolution into a model. Such stimuli are troublesome, incompressible: you can't wrap them into an economical, unified, higher-level conceptual category with zero prediction error and go on your way. You have to settle for lower-level representations of bits of the stimulus, and a kind of 'Warning: Failed to converge' at the higher conceptual level. The brain is nothing if not an economical beast. It doesn't want this kind of clutter hanging around. It needs to avoid it, ignore it, or tidy it up.

§

How does this relate to my experiences as an inter-discipliner, and particularly one who tries to bridge social science and biology? I have an interesting natural experiment to report here. For most of my career, I worked exclusively on humans, and not just any old humans, but (mostly) contemporary British humans. Some of my human work relates to phenomena like teenage pregnancy. I have a particular take on this phenomenon. I have argued that women from poor backgrounds who bear children at a young age are not necessarily 'making a mistake' or 'failing to exercise self-control'. They are making a 'contextually appropriate response'. That is, they are following a behavioural strategy that makes a lot of sense given their relatively short healthy life expectancies (in the poorest English neighbourhoods, women can expect to be in good health only until they are just over 50—why would they wait until they were 40 to start a family?); and given the modest economic returns to delaying childbearing when only low-skilled, low-paid jobs are available.[4]

My position is extremely congruent with other social-science perspectives. Social science scholars have also made the point that women who bear children young are not committing impulsive individual mistakes but responding, sometimes with some deliberation, to the circumstances in which they find themselves.[5] I'm making exactly the same argument, but I am prone to alluding to the evolutionary concepts of 'adaptive behaviour', 'lifetime reproductive success', 'fitness', and so on. I do this because evolutionary behavioural ecology, the source I drew inspiration from, provides rather useful general expectations (or methods for coming up with expectations) about how individual organisms should respond to their environments.

You might think, naïvely, that my teenage pregnancy work would fascinate and engage my social science colleagues, and make it easy to build academic bridges between social science and biology. I'm saying a lot of what you guys are saying, and then I am also relating it to a suite of more general concepts from behavioural biology that are already lined up and ready to be investigated. Isn't that exciting? I have given numerous talks of this kind to audiences from epidemiology, public health, sociology, and economics.

I have to tell you that it has not, in the main, gone very well (though there have been a few enjoyable exceptions). People have always been

---

4   Nettle, D. (2010). Dying young and living fast: variation in life history across English neighborhoods. *Behavioral Ecology* 21: 387–95, https://doi.org/10.1093/beheco/arp202

5   See for example Arai, L. (2009). *Teenage Pregnancy: The Making and Unmaking of a Problem* (Bristol: Policy Press).

very polite and friendly. The overt hostility to evolutionary approaches that people attribute to social scientists is not, in my experience, widespread. What I do experience is simply a slightly embarrassed awkwardness, and no return phone call. Paul Feyerabend said that scholars who introduce transgressive ideas find themselves faced, 'not with arguments, which they could most likely answer, but with an impenetrable stone wall of entrenched *reactions*'.[6] And the reaction in Britain is, most often, quiet puzzlement, a polite but slightly strained question or two, checking of the train timetable home, and perhaps sometimes—I am remarkably neurotic but, yes, I think it might sometimes be there—a fleeting moment of well-concealed *disgust*.[7]

I accept that my rather broad level of analysis simply may not answer the fine-grained questions that social scientists want to answer. But there may also be something deeper going on. Perhaps the central argument of my research on humans comes across as 'the thing that cannot be'. The central dogma of evolutionary biology is (allegedly) that genes make bodies and therefore the successful replication of genes is, in the final analysis, the determinant of what goes on. The central dogma of social science is (allegedly) the idea that context determines behaviour.[8] These two dogmas seem a long way apart, like you would need to choose one or the other. It may not be clear to my audience which one I have chosen; hence the reaction.

In fact, though, you don't need to choose. Let's start at the evolutionary biology end but move in the direction of social science: *because* the replication of genes is so important, and because the best way of surviving and reproducing is very different in different local environments, evolution has produced creatures that are highly sensitive to the contexts they get put in. Evolution, instead of making

6 Feyerabend, P. (2010 edition). *Against Method: Outline of an Anarchist Theory of Knowledge* (New York: Verso, p. 59).

7 The anthropologist Mary Douglas importantly linked the emotion of disgust to the violation of category boundaries, and hence, though she did not put it this way, to prediction error signals in the brain and the uncanny valley phenomenon: Douglas, M. (1966). *Purity and Danger: An Analysis of Concepts of Pollution and Taboo* (London: Routledge), https://doi.org/10.4324/9781315015811

8 'The idea that context determines behavior is the 'central dogma' of all social sciences from anthropology to sociology, political science, psychology and economics.' Glass, T. A. and U. Bilal. (2016). Are neighborhoods causal? Complications arising from the 'stickiness' of ZNA. *Social Science and Medicine* 166: 244–53, https://doi.org/10.1016/j.socscimed.2016.01.001

a train that can only go down fixed tracks, has made a self-driving autonomous vehicle that can go to a wide range of places according to the landscape it finds itself on. Genes have done this, if you will, because it is in ultimately in their replicatory interest. Thus, the sensitivity to context our genes give us is not a random one, but one structured toward certain needs or goals. On the other hand, it's not as though all the possible consequences that could emerge once a whole fleet of self-driving autonomous vehicles start driving around a town were already present in the heads of the engineers who designed the vehicles; of course they weren't. Likewise with genes. So you don't need to choose between genes and context, any more than you need to choose between brake pads and traffic jams.

You might think this kind of position would make everyone happy, and we could all get on famously. But all too often, it seems to fall into the uncanny valley. What does this guy really think? He talks like a social scientist about these contextual factors, then he starts mentioning genetic fitness, and it is as if we suddenly see his battery pack and realize he is really a replicant. If we stick to his actual claims about context—for example that poverty and emotional trauma predict teenage pregnancy—then we already knew those facts, and don't feel any great need for any explanation for them beyond those we already have (indeed, these facts *constitute* a certain kind of explanation for the behaviour). If we focus on all that evolutionary baggage, then we end up with something that gives a rather large prediction error signal whether you try to think of it as a duck *or* as a rabbit. So the view that social science and evolutionary biology can be productively integrated into a synthetic position incorporating information from both is hard to hang on to.

The uncanny valley is steep on both sides. The same talks that produce polite puzzlement in social science departments produce just as much puzzlement, or perhaps even more puzzlement, in Zoology departments. A recent survey of working biologists found only 60% agreeing that what we learn from humans is relevant to understanding other evolved creatures, and the survey probably over-sampled biologists working on humans.[9] The explicit reasons evolutionary

---

9   Briga, M. et al. (2017). What have humans done for evolutionary biology? Contributions from genes to populations. Proceedings of the Royal Society B:

biologists give for a queasiness about humans vary from the sensible and practical—like the long generation time and the difficulty of performing true experiments—to the completely bizarre and question-begging—like 'humans are influenced by social factors'.

§

Five years ago now, I began to work on European starlings (sensitivity to context thereof, as it happens; the starling is 75 grams of very sophisticated autonomous vehicle). One consequence of this shift, unsurprisingly, has been greater interest in my work from zoologists and evolutionary biologists. This is the payoff for having climbed out of the uncanny valley up the biological escarpment, and hence generating no troublesome error signals from the fact that my study species wears clothes and watches cable TV.

Strikingly, colleagues from the social sciences and humanities *also* engage with my starling work much more enthusiastically than they do with my human work. *You're that starling guy who shows that what you get to eat early in life affects your behaviour when you grow up. How fascinating! I* love *what you do! Actually, I was watching the starlings in my garden, and I was wondering....* I am continuously pumped with feathery questions, questions that one might quite comfortably ask of an exotic human society (and to which I usually do not know the answer). This curiosity extends to the general public too. When we are working in the field, people stop their cars to ask what we are doing, why the starling has become so much rarer, whether individual birds use the same nest box every year, whether starlings can feel pain, and whether it is true that the male starling must offer a nuptial gift of aromatic greenery, placed ceremoniously on the nest, before the female will begin to lay.[10]

There is no unease or edge in any of this questioning, just delight. People grasp that there is a form of life over on the other side of the uncanny valley in birdland, and they know that they don't know what it is like. Hence they love meeting someone who has tried to go there and can talk to them about it. That person generates no prediction error signal and poses no embarrassment. Even the insight that the form of life

---

Biological Sciences 284: 0171164, https://doi.org/10.1098/rspb.2017.1164
10   It is true.

in birdland has features recognisably akin to our own—parenthood and nuptial gifts—is just fine, as long as we are talking about two *different* worlds with some parallels, not a mixed world. The two sides of the uncanny valley can echo each other's landscapes in a kind of aesthetic way, and each can contemplate the other admiringly from afar. Waving from the opposite hillside is a lot easier than living down in the uncanny valley.

§

The same processes that give us the uncanny valley produce a continual loss of the nuanced middle ground in the behavioural sciences. I am prepared to bet that despite my rather careful explanations about autonomous vehicles, the compatibility of genes with sensitivity to context, and so forth, there have been times when someone has said: 'We had that Nettle here yesterday. He believes teenage pregnancy is caused by genes rather than the environment!'. And it is not just me that has this problem. Mischaracterization of evolutionary approaches to human mind and behaviour—particularly, the claim that such approaches must deny the importance of context—is pervasive, despite repeated and explicit statements to the contrary by the proponents of these approaches.[11] One has to ask oneself: is it our fault for not being clear, their fault for not listening, or is something more general going on?

Failure of scientists to correctly represent one another's positions is not surprising. Classic work by Frederick Bartlett showed that if you give someone an irreducibly complex shape, have them copy it, than have a second person copy the first, a third person copy the second and so on, the shape soon becomes less complex, and closer to something simply nameable like a cartoon cat or a letter 'A'. Monica Tamariz and Simon Kirby showed that all you need to do to make this loss of nuance happen is to have each participant need to store the shape in memory for a short time.[12] It's the compressive processes of cognition—specifically, to be

---

11 See Kurzban, R. and M. G. Haselton. (2010). *Making hay out of straw: Real and imagined controversies in evolutionary psychology. In Missing the Revolution: Darwinism for Social Scientists,* (J. Barkow ed., Oxford: Oxford University Press, p. 149–66), https://doi.org/10.1093/acprof:oso/9780195130027.003.0005

12 Tamariz, M. and S. Kirby. (2015). Culture: Copying, compression, and conventionality. *Cognitive Science* 39: 171–83, https://doi.org/10.1111/cogs.12144

stored, something has to be expressed in terms of an internal model—
that inexorably drive the distortion and polarisation of complex ideas.

In another revealing experiment, participants were trained on a
mathematical function: they saw one x-value at a time, represented as
the width of a bar, and they had to select a corresponding y value.[13]
They were given feedback until they got it right. They were then given a
set of test trials with no feedback: on each trial, the x was given, and the
participant proposed a corresponding value for y, which was recorded.
The next participant was then trained, not on the original function's x-y
pairings, but on x-y pairings that the previous participant had offered
during their test trials; their version of the function.

The results are some of the most remarkable I have ever seen.
Within a depth of about seven participants, all the functions had
become positive linear ones, regardless of what the starting point was.
Curvilinear functions became positive linear ones. Randomly generated
patterns became positive linear functions. Remarkably, even negative
linear functions became positive linear functions. In short, within a
few rememberings and retellings, the image we were left with carried
no information at all about the stimulus we started out with. It carried
information *only* about the kind of pre-existing schemas that people
find easiest to learn and remember. This is a very sobering result, given
that the ideas worth devoting one's life to are nuanced, layered, and
don't fit into convenient pigeon-holes like nature or nurture, genes or
environment, individual or society.

§

What happens to ideas that don't fit neatly into any existing schematic
paradigm? Mostly, they get turned, like the functions in the function-
learning experiment, into something more black-and-white, and less
accommodating. (This can give an airtime advantage to ideas that are
black-and-white and not very accommodating to start with.) Occasionally,
though, a new idea gets a foothold. In effect, the community's internal
models get updated enough for there to be a new recognisable category

13   See Griffiths, T. L., M. L. Kalish and S. Lewandowsky. (2008). Theoretical and
     empirical evidence for the impact of inductive biases on cultural evolution.
     Philosophical Transactions of the Royal Society B: *Biological Sciences* 363: 3503–14,
     https://doi.org/10.1098/rstb.2008.0146

of argument out there. The idea becomes sufficiently stabilised to hold its identity. A new explanatory schema is born.

In my main field, human evolutionary behavioural science, there were some examples of this back in the 1980s and 1990s. The new explanatory schemas included, for example, the idea that our current minds (and bodies) are specialised for life in small-scale Pleistocene societies, not the kinds of societies we live in today; and the idea that culture is an inheritance system, parallel to genes. The emergence of these schemas gave rise to a process of fragmentation as groups of researchers coalesced around one or other of them, forming the mini-disciplines of 'evolutionary psychology', 'cultural evolution' and 'human behavioural ecology'.[14]

Once new schemas such as these get a foothold in the middle of the uncanny valley, what happens next is predictable but somewhat ironic. The schemas attract paradigmatic adherents, who are often more dogmatic than the founders, and the adherents form cliques with one another. This shows us that most people are unable or unwilling to live out there in the blazing sun of complex and ambiguous phenomena with just their bodies and their native wits to protect them. They crave the epistemic shade provided by a micro-community with a nameable paradigm: a comforting system of assumptions, sacred texts and fellow worshippers. They crave this for their own cognitive ease, but also because of the social processes involved. It's easier at an academic party to say 'I am an X' than 'I look at A with a bit of Y and a bit of Z'.

So once a mighty tree takes root out there in the uncanny valley, soon there's a little copse of saplings underneath. And these copses, these mini-disciplines, come to have some of the unfortunate properties of the big disciplines they initially hoped to bridge. They have mini-uncanny valleys all around them, between them and neighbouring copses. They acquire purity to be defended. The boundaries of the mini-disciplines are in some ways more troublesome than the macro-boundaries like the social science/biology boundary. They are troublesome because the copses, though verdant, are small enough to have limited resources and odd founder effects. The adherents defend their copse. They try to

---

14 See Sear, R., D. Lawson and T. E. Dickins. (2007). Synthesis in the Human Evolutionary Behavioural Sciences. *Journal of Evolutionary Psychology* 5: 3–28, https://doi.org/10.1556/jep.2007.1019 for a review.

suppress competitors, and the competitors you most need to suppress are those who encroach most closely on your niche. I have been struck, as I have watched these mini-disciplines take root in my community, how easily the new adherents abandon the responsibility of pluralism. They don't often feel the need to visit the other copses, except maybe to dismiss them. They don't even engage much in an ongoing way with the broader intellectual sources—ethology, cognitive science, social science—from which their inter-disciplinary mini-disciplines sprang, and on which its future flourishing depends.

All this means trouble if you want to operate in the uncanny valley between evolutionary biology and social science, but *don't* want to commit exclusively to one of these copses. (You may see the value in all of them, but also recognize the incompleteness of all.) On the macro-scale of engaging distant colleagues, it's hopeless. It's hard enough to get them to understand that there is *one* way of combining aspects of evolutionary biology and aspects of social science. To get them understand that there are *several different* ways of combining them, and these are not interchangeable, that's hard. One risks, as it were: 'I've heard about one attempt to be evolutionary about modern humans and I didn't like that, so I assume you're just the same' (and if not that, then: 'you evolutionary people don't even agree amongst yourselves'— it's hard to win at this game). Within the tiny community of human evolutionary behavioural scientists, too, it's hard to be an in-betweener. People need you to either be one of their own copse's flag-bearers, or else a straw man. You can get castigated for deviations from schemas you never intended to adopt. Many of the most interesting empirical findings—observations that would extend all of the copses, but are not immediately recognisable as central exemplars of any of them—simply languish. They fill badly needed gaps in the literature.

§

The only thing worse than having people not cite your work is having them cite it. When your work does get picked up in the literature, it's salutary to look carefully and try to identify the claims that those citations are used to support. When I do this exercise, what I often see is that those claims are not really the claims I made; they are *somewhat* similar claims that are either more familiar, or more obviously ridiculous.

The argument that Ian Rickard, Willem Frankenhuis and I have been developing about why childhood family conditions have an effect on a wide range of adult outcomes in humans *really isn't quite* the argument that the childhood household furnishes cues about the harshness of the adult environment (and it really isn't quite the argument that childhood stress just messes up your brain, either).[15] I really have *never claimed* that the reason some people behave less pro-socially and more anti-socially than others is because they are following a 'fast life-history strategy'.[16] And my favourite recent example: Melissa Bateson, Clare Andrews and I wrote a paper giving an evolutionary take on human obesity.[17] The first draft started with a long background section in which we were very critical of the widespread idea that contemporary humans are obese because fats and sugars were rare in ancestral environments, and thus we have not evolved control mechanisms for saying 'stop' when these things are abundant. (There are numerous problems with this idea, at least in its simplest form, for example that many humans live in affluent societies and never become obese). That background section didn't make the final edit, because that idea was not the main point of our paper anyway. Within six months of the paper being out, guess what I saw in a draft manuscript I was reviewing? 'Sugars and fats were rare in Pleistocene environments, and so humans have not evolved restraint mechanisms to stop them over-eating when these are available (Nettle, Andrews and Bateson 2017).' If you didn't laugh, you would probably cry.

My case is not unusual or worthy of any special consideration. It's just the one on which I have the richest data. It illustrates a more general problem: what people actually say is not what we remember them as having said; and not what it would be more convenient from the point of view of our agendas if they had said. There are two commonplaces from the history of science that fit with these observations. The first is

15 Rickard, I. J., W. E. Frankenhuis and D. Nettle. (2014). Why are childhood family factors associated with timing of maturation? A role for internal prediction. *Perspectives on Psychological Science* 9: 3–15, https://doi.org/10.1177/1745691613513467

16 One of the places I really don't say this is in: Nettle, D., A. Colléony and M. Cockerill. (2011). Variation in cooperative behaviour within a single city. *PLoS ONE* 6: e26922, https://doi.org/10.1371/journal.pone.0026922

17 Nettle, D., C. Andrews and M. Bateson. (2017). Food insecurity as a driver of obesity in humans: The insurance hypothesis. *Behavioral and Brain Sciences* 40: e105, https://doi.org/10.1017/s0140525x16000947

that the troublesome data that will eventually necessitate a change of scientific view often exist in plain sight for decades or even centuries before the change of view happens. The problem is not that the critical observations have not been made. The problem is that the community does not know where to put them in its mental models, so it either ignores them, or misrepresents them as something different from what they really are.

The other commonplace is that new ideas get dismissed as wrong until exactly the point where people say that they are obvious and they always knew them anyway. From the point of view of mental models and prediction errors, we can see what is happening here. Initially, I hear idea X one or two times. I can't assign it to any mental model. It just makes error signal. It must be wrong. But later, I have heard it dozens of times, enough for it to have changed the representational options available in my internal mental model. Here's that thing X again! I've already got a mental model of it, so it seems obvious!

§

The human brain is often described as one of the greatest remaining scientific problems. I think this is true, and not just in the way it is usually meant. One lesson for researchers is the need to be extremely clear and do a lot of very patient, cheerful, and if necessary, repetitive signposting. Some of our most successful conceptual innovators have been prepared to do this, year in year out, writing the same paper for different audiences, or even the same paper for the same audience, until the penny begins to drop and the idea gets recognised for at least approximately what it is. If like me you are prone to constant shifts of views, banging the drum for the same idea year after year is not something that comes easily. Clear repetition also hardens all too easily into dogmatism and parochialism. Nonetheless, some modest insistence is often necessary.

I would like to close by proposing a scientific innovation: the anti-abstract. This is a short summary of what a paper does not say. I think all papers should have these, published immediately after the conventional abstract. Casual readers could read the paper, spend 30 seconds counting backwards in 7s from 116, and then read the anti-abstract. If they get a big error signal from comparing the anti-abstract to their memory of what the paper said, then they know they need to read the paper again, paying closer attention. They need to put the effort into creating a finer-grained representation of the paper's claims: those claims are not what they thought at first pass. The anti-abstract would be very handy as a one-stop-shop for all the likely peer reviewer misinterpretations. Instead of having to take paragraphs of precious Introduction and Discussion laboriously setting out a load of ideas that are not in fact the ideas you want to test, you could simply mention them in the anti-abstract as claims you are neither advancing nor even considering, and for which your work should not in any circumstances be mistaken.

I am really looking forward to writing anti-abstracts. In fact I might start doing so, and keep them in the file drawer for the day academic journals start asking for them. I can imagine beginning with the broad theoretical anti-sweep: 'Researchers have argued that individual differences in many behaviours can be mapped onto a single underlying continuum of fast versus slow life history strategies. This paper is not an exemplar of those arguments'. Then there's the anti-summary of methods: 'Methods we did not use in this study include the public goods game'. And of course, the anti-implications: 'Results are not interpreted in terms of the poor lacking self-control'; or 'Our findings do not imply that fertility decisions are controlled by specific genes'. The best thing of all about the anti-abstract is that it gives the perfect chapter and verse defence when you get mischaracterized. Stronger as a defence than 'I never said that' is 'Look, I actually anti-said that in the anti-abstract'.

Undermining misapprehensions at source is surely a worthwhile goal. Mutual misconceptions comfort and simplify — the inside of one's prejudices is an easy place to live, after all — but are not, in the end, very useful. The more we clear out misconceptions about what other groups are saying, the more connected our conversations might become to the world itself.

13. Staying in the game

> 'I've done this a long time', she said.
> 'I've seen long careers and careers cut short.
> The difference is how you handle the
> darkness'.
>
> –Michael Connelly, *The Late Show*

A student from another university interviewed me. An assignment for one of her classes was to interview a researcher from a field of her choice about their work and their career. She chose me. As one of her prepared questions, she asked me what my aspirations were for the next ten years. I replied that, looked at in the round, what I hoped for in ten years was to have remained alive, and ideally continued doing some research.

I think the student was a little taken aback. Perhaps she had been told how focussed and driven successful researchers have to be; how they have to have boundless confidence and a nine-point plan. So to hear that the best I hoped for was merely to somehow stay in the game (and that I was not completely confident I would succeed in doing so) was disarming to her. In some ways, though, my answer was a sensible one. Science (and other creative endeavours) are rather like animal life in a Darwinian world. There are many more ways to be dead than alive, and the vast majority of all lineages die out. To succeed—more exactly, to not yet have failed—is to still be in the game; to flourish is simply to not yet be extinct.

A document has done the rounds at my university that attempts to inform young researchers about the career structure of academia. At the left-hand side is a fat horizontal arrow, pointing rightwards, which is marked 'PhD student'. Then to the right of that one, pointing in the same direction, a slimmer arrow is marked 'Post-doc'. To the right of post-doc, and about half the girth, is the arrow marked 'Independent research fellow'. Finally, at the extreme right is the slimmest arrow of all 'tenured university professor'. It is less than a quarter the width of

 https://doi.org/10.11647/OBP.0155.13

the 'PhD student' arrow. And where each horizontal arrow meets the next, there is a vertical downwards arrow reminiscent of the way sinks or earths are depicted in physical diagrams. 'Exit academia', this arrow is labelled.

You might think if you make it to the right-hand side, to 'tenured university professor' then the shoals have been navigated and you are in the game forever. Not really. The research longevity even of professors is finite. There comes a point where things get becalmed. What was front page news when they were a PhD student only makes the middle pages now. They have friendly but sad annual meetings with the Dean, like the solid lay wardens of a settled religion whose congregations are declining; wondering if the fervour will ever quite come back. They are gently asked, or volunteer themselves, to take on senior administrative functions, or more teaching. There is nothing wrong with this, of course. The life of the body has to be lived. Besides, these functions are important ones. Still, they are not what most of us went into it for. We were seduced by the primary research process: the idea that you could find a question; hit on your own approach; perform and manufacture the work; and finally, see it there in print, with your name attached, a thread woven in to the tapestry of human knowledge. A thread of memory.

Perhaps the student was seeking some pearls of wisdom about how to stay in the research game. I am not sure I managed to give her any; and my immediate thought was that I am the last person she should have been seeking to emulate. If she knew how narrowly I have hung on, I thought, she might have chosen someone else for her assignment. But, on second thought, maybe the people who have narrowly hung on are the most informative. After all, you can find any number of books about the practices of the most successful, the mega-stars—the case study of the 'winner' is an established genre. It has always struck me that, interesting as these books might be, in a way they pose the wrong question. It's very hard to win the big prizes, and thus of some interest to know how the few people who do, do it. But the more pressing question is: how do you stay in the game *without* winning them? How can you live a worthwhile and satisfactory life if you are a competent businessman but not a Bill Gates; a competent actor who is not a Marlon Brando; or a useful scientist who never garners the accolades of a Stephen Hawking? *That's* what takes real grit, humanity, wisdom and technique: to just be there, quietly, purposefully, usefully, afflicted with neither pomposity nor despair, whatever the weather. To flourish in the middling state. The character we respond to most in Ronald Harwood's play *The Dresser* is not 'Sir', the famous lead actor, but Thornton, the long-term bit-part player who loves his craft and stays in the ensemble, taking small roles with dignity year after year. A good life: always working, always touring; but never London.

I suppose, then, I am well qualified to say something about staying in the game. I have done it for over twenty years. I have contributed to research on several topics in biology and the social sciences. I hold a full professorship in a decent research-intensive university. I get some very nice academic invitations. So it seems I have avoided the 'exit academia' arrows, possibly even with aplomb. But—and this is what qualifies me—I haven't avoided them nearly as easily as you probably think. If it looks that way from the outside, that's only because you lack the data I have. You don't see the failures, the false starts, the wasted time, or the awkward conversations with the Dean. (All of these are ubiquitous, by the way, just not well written up in the literature.) And opposite the Scylla of the 'exit academia' arrows has always lurked the Charybdis of my own demoralization: walking away from the game even if the game does not eject me. I gave up completely once, for several years, then eventually clawed my way back in; a second time I partly gave up but left myself attached by a lifeline, a lifeline I duly climbed up within a year or so; and probably two or three more times over the years I reached the point of starting to make other plans. Periodic demoralization and depression are not rare amongst researchers. It's not 'not caring any more', or 'not being able to be bothered', as depression is often and erroneously characterized. It is caring *so* much, being *so* bothered, that one cannot advance on any front. One drowns in one's own disorganized and gradually souring passion.

There's a lot to trigger demoralization an depression in the typical diet of the contemporary researcher: intrinsic uncertainty about the subject matter and one's progress through it; rewards that are always deferred and whose arrival is highly unpredictable in time; structurally frequent rejection that is hard not to take personally; permanent opportunities for unfavourable social comparison; and, at least in British universities since they became so obsessed with research 'metrics', an officially deniable but still palpable sense of threat. Despite all this, though, I still believe that to spend one's working life as a weaver on the collective tapestry of human ideas is a noble calling, and a privilege. You've just got to find a way of doing it that suits you, works well enough, and keeps your spirits up most of the time. So, for what it's worth, I thought I would set down some of the lessons I have learned trying to stay in the game all these years. These are the lessons that I wish I had told that student about, and I wish she had asked me for.

Lesson 1. Every day has to count for something

I try to start each working day with a period of uninterrupted work. Work, for me, is: collecting data, analysing data, writing code, drafting a paper, writing ideas in a notebook, or just thinking. Things that do not qualify as work are: background reading, literature searches, answering correspondence, marking students' assignments, peer-reviewing a paper, sorting out my website, correcting proofs, filling in forms, tidying datasheets, having meetings, and so on. These are work-related activities, which are necessary for work to be possible, but are not the work itself. It's very important that this distinction be maintained. Don't try to do a simultaneous mixture of the two (it's obvious how that is going to end); and *always* do work first, work-related activities second. For example, I might decide to start the day with two hours of work, and then, at 11 o'clock say, allow myself to start the work-related activities required to keep the show on the road. Working requires emotional commitment. It needs to be planned the day before. The phone is off; the email is off; if you are likely to be disturbed, there is a 'do not disturb' sign on the door. And if like me your work time is first thing in the day, then you never peek at your email before starting. I like to be very quiet in the morning, not even getting into too animated a discussion, reserving my energy for doing some work; then, once work is done, I can be more relaxed and expansive.

I thought that starting the day with a period of work was just a system that I had discovered by trial and error works for me, but something very like it turns out to recur in descriptions of the creative life,[1] and self-help

---

1   For example, Murakami, H. (2008). *What I Talk About When I Talk About Running* (New York: Knopf); Hardy, G. H. (1940). *A Mathematician's Apology* (Cambridge: Cambridge University Press).

books on how to be more productive.[2] And these accounts stress that it is imperative to do work *every day*. I don't mean you shouldn't take days off: I don't usually work at the weekends, and I take about a month completely off every year. I mean that if today is a working day, it must contain a period of *work*; it cannot be completely filled up with work-related activities. Though there are some challenging exceptions (e.g. field work, travel), I try to maintain the every-day rule. If your day is filled with meetings, then fine; you have to get up an hour earlier that day to at least get one hour in on that particular day. And when it comes to scheduling meetings, you will probably have some latitude; don't schedule them for 9 or 10am. If people ask to see you, say you are not available at those times, but that you are happy to see them at lunchtime or afterwards. And it's good, when you are thinking of whether to drop into a colleague's office, to be considerate about the time of day: are they likely to be working, or just doing work-related activities? Could this be sorted out over lunch or in the afternoon?

Why is the every-day rule so important? Well, there are only about 200 working days a year, so 1 day is 0.5% of a year. Proper work is really hard. And we are lazy, weak creatures. We are not set up to forage in really hard ways when much easier ways of foraging are only a click of the browser button away; why would we be? If you allow that there is some set of circumstances X that permit starting the day without settling to proper work first, then you will manage to convince yourself that X obtains quite often: namely, whenever you are a bit tired or stressed; when the problem you are working on is getting difficult; or when your belief in your current project is a bit insecure. But not working on it today won't solve any of these problems; indeed, will make them worse. One day without useful work rapidly becomes two or three, and then a whole week; then before you know it, your working practice has descended into undifferentiated low-value grazing on work-related activities, without really getting anywhere. That's why the difference between amateur writers and professionals is that amateur writers write when they feel inspired, whilst professionals write every day. See your designated work time out each day, even if it means staring at the wall for an hour.

---

2   For example, Newport, C. (2016). *Deep Work: Rules for Focused Success In A Distracted World* (New York: Grand Central).

You may well say, it's all very well for you to advocate this idyllic lifestyle, since you work at a nice university that gives you low teaching and administration burdens. True enough, but I make two observations. The first is that I know plenty of people who have lower teaching and administration burdens even than me, and still don't get much done. The second is that your deep work doesn't need to, and probably can't, take many hours out of each day. Even one good hour per day would cumulate quite rapidly over the course of weeks and months. And surely you can carve out one hour per day? In fact, I find few historical examples where real work goes on for more than a few hours per day, even absent any other demands. Murakami writes in the morning and spends his afternoons training for his marathons, and the great G. H. Hardy would work for 3-5 hours on his mathematics, then take himself off to Fenners to watch the cricket. For those of you who enjoy palaeo-bullshit, 3-5 hours a day was Marshall Sahlins' suggestion for how much time humans spent working (i.e. foraging) in hunter-gatherer societies. The remainder was available for rest, social life, self-maintenance, and just being. This pattern, Sahlins teaches us, constitutes a form of affluence; not the affluence of the consumer society, but the affluence of doing a bit of the stuff that matters most deeply to you, and having simple wants beyond that.[3] Once you have done your 3-5 hours, there is time for really talking unhurriedly to the people you work with; going to talks; having walks; understanding how to do something you don't currently understand; or whatever.

Daily deep work keeps the black dog away, for there is nothing worse for mood than the sense that one is not progressing. And it can spiral in a bad way: the more you feel you are not progressing, the worse you feel; the worse you feel the more your hours become non-deep junk; and the more exhausted you are by non-deep junk hours, the less you progress. As they say up here, *many a mickle makes a muckle*. This translates roughly as: large things are composed of many small things. This is not a commitment of Northern folk to a particular kind of reductionism. Rather, in this context, it means that the biggest gains

---

3   Sahlins, M. (1968). *Notes on the original affluent society. In Man the Hunter*. (R. B. Lee and I. DeVore eds., New York: Aldine Publishing Company, p. 85–89); Sahlins, M. (2009). Hunter-gatherers: Insights from a golden affluent age. *Pacific Ecologist Winter* 2009: 3–8, downloadable from: https://pacificecologist.org/archive/18/pe18-hunter-gatherers.pdf

to your overall productivity stem not from any macro-level great leap forward, but from small changes to your daily practice. If you make each and every day a bit more productive, then the months and years kind of take care of themselves. And if, as one colleague complained to me a while ago, you are putting in eighty hour weeks and still not getting your important goals achieved, then the answer is not to put in more hours: it is to put in fewer.

Commitment to deep work entails choices. It means not travelling to more than one or two conferences or workshops a year; not taking on peripheral involvement in collaborations extraneous to your main purpose; not filling up your diary with 'it might be useful...' training courses or committees; or applying for money you don't really want or need. Saying no is hard; we worry about missing out on something, and about the social or reputational consequences of a refusal. In my experience, these are siren voices. It is better to be a polite 'no' than a 'yes' who turns out to be over-busy, late, and frustrated; better no application than one rushed together at the last minute. A phrase I find useful is 'whilst I would love to, I do not have the capacity'. I enjoy its ambiguity, and it never leads to anything other than a sympathetic and understanding return message.

Lesson 2. Cultivate modest expectations

A friend of mine talked to me years ago about how to start a theatre company. Hire a small cheap room and invite three people along, she said, then spend your time making work that means something to you. If two of the three people come along, that's fine. Don't sweat too much about getting West End producers to attend. As a foolish young man, I thought this was rather negative. What is wrong with these arts people? What's the point of making work if hardly any one sees it? But now I am older I understand her wisdom. If you can manage not to care who comes, you can make the work with freedom and right mind. If you are worrying about whether the West End producers will show up (they probably won't), then you can't. And if your expectations are modest, you can not only meet, but sometimes exceed, them.

I was at a discussion meeting recently where a number of us from similar fields had been assembled. The organizer said at the outset that he thought together we could get a paper into the journal *Science* (arguably the world's most prestigious journal) from the discussions of the meeting. I asked what the paper would be about, and he replied that he didn't know yet but he hoped it would become clear over the course of the two days. This, it seems to me, is the very antithesis of my theatre friend's wisdom; it is focussing on the accolade and not what the accolade is an accolade for. The consequence is not just that we don't have a paper in *Science*. We don't have a paper at all.

I have never hit the heights of papers in *Nature* and *Science*. For a long time, this and other failures rankled. I know that it cost me at least one job, and there is the slightly uncomfortable feeling of being thought of as never having quite made the grade. There are two possible philosophies here: *keep on trying, winners never give up*, or *find an inner sense of value in your work, rather than relying on glittering prizes*. I respect and can see the logic of both philosophies, and people are very different in terms of what keeps them going. But I would be a card-carrying hippy if hippies weren't so against the carrying of cards, so I think you can probably guess which one temperamentally attracts me.

Why? Well, the glittering prizes we academics strive for are positional goods kept deliberately scarce by bureaucratic or commercial interests, and allocated in ways whose relationship to long-term value is probably quite weak. For example, *Nature* is a for-profit enterprise that rejects nearly everything in order to defend its exclusive market position. If we all send everything there, the rejection rate goes up. If we all increase the quality of our science, it still nearly all gets rejected, by the very design of the institution. The idea that all good papers can be in *Nature* or *Science* is as ludicrous as the idea that all Olympic athletes can get gold medals, but without the strong link between actual ability and finishing position that obtains in the Olympics.

If you are in the habit of comparing yourself relative to peers or rivals against simple external yardsticks, your modal experience will be a feeling of failure (believe me, I've been there). For a start, the external comparators available to us all have right-skewed or 'winner take all' distributions. In such distributions, the median is always well below the mean. So most people look worse than average, and all but one person can find someone who is doing *a lot* better than them. Besides, there is an unhelpful asymmetry of information. You get new information on your own progress every day (have I published any new papers since yesterday? No.). For your peers and rivals, you check their website maybe once a year (*six* new papers since I last looked!). So of course it looks like they are progressing better than you. But they may not be: you are just sampling less frequently for them than you are for yourself. The solution? How about trying not to think about it?

There is a subtle issue here of the allocation of mental energy. My theatre friend's modesty of aspiration came from a deep understanding that mental energy allocated to chasing the external trappings of success is not being allocated to the authenticity of the work. By trying to make a West End hit, you make something which is, at best, derivative of previous West End hits. Its capacity to be truly transformational is probably limited. Great art often begins on the fringe. Similarly, valuable future paradigms and innovative ideas start life in obscure places. Journal editors cannot yet see their potential, and the authors themselves are tentatively feeling their way into something new. So by focussing on capturing the established indicators of prestige, you distort the process away from answering the question that interests you in an authentic way, and into a kind of grubby strategizing. Or so I tell myself, admittedly through clenched teeth at times.

In truth, the best things in my career have been fringe efforts, done in ludic spirit with no funding, and published in journals that base publication decisions on ethical and analytical soundness, not some editor's hunch about whether something is a West End hit or not. This has allowed me licence to do what I thought was interesting, even though the big journals or regular funders would never look at them. Whether they stand the test of time, we won't know until well after I am gone; but that would have been true too had they been published in *Science* or *Nature*. The outputs of which I am most proud are judged pretty much worthless in terms of the metrics like journal impact factor that universities obsess about. This hurts. However, I am consoled by the fact that there is a small band of people, spread across the world, who really get what I was trying to do, and think it is interesting. I know because they write to me; because they actually download and cite those papers a fair amount; and because as I get older I start to see tiny signs of my influence in their work. That's the best I can hope for, and all I try to need.

Of course, your Dean might have a few words to say about your Stoic disavowal of impact factors, massive grant income, and what not. This is a difficult problem. It is important to understand that your Dean is not a bad person, or anti-intellectual; they are just relaying the pressures that are hitting them, on down the line. And they are right that our positions should not be sinecures: students and tax-payers pay our salaries, and are entitled to audit what they are getting. You should not go out of your way to *avoid* prestigious publications or grant income when the opportunity arises. It's just that somehow those things have not to dictate your direction or self-worth; and you have to find a way of keeping going whether or not they come. I suppose I have had a happy knack of paying enough unto Caesar as to keep the Roman army off my back, but not so much that I lost my independence of spirit. A life I admire is that of Spinoza, who preferred to work a fraction of his time as a lens grinder than accept patronage or a university chair. This meant he could stay in the game on the strength of his lenses (which by all accounts were very fine), and pursue his philosophy with complete freedom and honesty. I am no lens grinder, but I try to pay my way in the world through a very steady flow of openly shared, thoughtful, workman-like science, even if most of it is not deemed stellar; trying to be a public communicator; being a good-enough teacher; and contributing my fair share to the common weal of university life.

Lesson 3. Publish steadily

I've been able to stay in the game despite not hitting the big metrics because I have always managed to publish one or two workman-like empirical papers every year, pretty much without exception. I have often done other stuff too: popular books, a textbook, reviews, more speculative ideas pieces and so on. But I do not *depend* on these other things. Every year, whether or not these other things happen, there is a peer-reviewed primary paper or two, not just with my name on, but actually written by me, with empirical data or original computation reported in it. This has been important both for avoiding the 'exit academia' arrows, and for keeping depression away.

I think the mistake a lot of people make is focussing too much on getting the big shot, the single career-establishing paper in a top journal, and therefore not quietly building up a solid, progressive portfolio of sound work. Think of staying in the game as trying to keep your head above water. You can achieve this by giving a couple of small kicks with your feet per minute. Each of these only imparts a limited amount of energy, so if you follow this strategy, you can't afford to miss a minute. If you do, you will start to sink, and your small kicks may not be enough to regain the surface. On the other hand, as long as you keep your small kicks regular enough, they will keep you smiling indefinitely. An alternative strategy is to come up with a super-duper big kick that will send you free of the surface for many hours. Good for you if you manage this, but on average, your attempts will fail (all mine have). So you spend a minute trying to devise your schmancy big kick, and during that minute, you haven't produced any normal small kicks. That means you are a bit lower in the water, and so that big kick is going to have to be even bigger (hence, even less likely to succeed) when it comes. So you get even more focussed on making your big kick *really* big; this is hard, and absorbs all your attention, and another minute goes by in which your position in the water has declined very slightly. The danger is, of course, that in the end you reach the point where the kick you need to save yourself would be infinitely large.[4]

I think my great strength is that I have always continued to produce something moderately useful, even when things weren't going well, and even when the big, bold, transformative ideas I so hanker for have eluded me. This has kept my head comfortably above water—indeed, left time and energy to strive after other things too—whilst I watched cleverer people than me gradually disappear into the 'in prep' section of their CVs, never to return. Relatedly, although I have made many false forays on my journey, I have got *something* out of every foray I have made, be it a methods paper, a model, a review, or a minor empirical study. This is a good knack. So if you are worrying about staying the game, rather than planning your next *Science* publication, I would ask yourself where your 1-2 solid papers each year are going to come from.

---

4   Devotees of foraging theory may recognise the spirit of David Stephens' classic risk-sensitive foraging model here: Stephens, D. W. (1981). The logic of risk-sensitive foraging preferences. *Animal Behaviour* 29: 628–9, https://doi.org/10.1016/s0003-3472(81)80128-5. And there are also echoes of Dean Keith Simonton's work showing that more successful creative people—whether in academia or the arts—are distinguished from less successful people mainly by producing more stuff overall, even though much of it is minor. See: Simonton, D. K. (1997). *Genius and Creativity: Selected Papers* (Greenwich, CT: Ablex Publishing).

Just as you should not go a single day without proper work, you should not go a single year without publishing anything, as one year rapidly becomes three.

Lesson 4: Get your hands dirty

Some people go into research because they enjoy the technical stuff: building and manipulating equipment, designing and carrying out experiments, being in the field, and so on. The problem these people have is that they under-invest in writing up everything they have done. They end up with mounds of unpublished data and cool techniques that have not really led to concrete outputs, and hence have not contributed to the field in the way they should. Other people conceive of the job of researcher as closer to the job of novelist. I don't mean they make everything up. I mean that what excites them is the writing, the putting it all together into a text that manages to capture their varied manifold of ideas and observations satisfactorily. I am the writer type: what has always attracted me is authoring wide-ranging books, articles, syntheses. Laying out the big ideas. I am never happier than when I have a free morning with a laptop and a pot of tea.

Whereas the technical type person only writes up as a last resort, to get the next round of funding or whatever, the problem with writers types like me is that we spend *too much* of our energy on the writing up. We view the gathering data as no more than background research, assembly of exemplary material for the writing we are doing. As a consequence we pick fruit that hangs too low: we end up using poor easy methods like online surveys, doing hasty secondary analyses of existing data, or just giving up the pretence and writing purely verbal papers that make various assertions in a sometimes appealing but often rather approximate manner. The only regret I have about my chequered career is that I have spent a bit too much energy on writing up — reviews, discursive papers — and not quite enough challenging myself by getting my hands dirty with primary research.

It's not that I don't see the value of the verbal argument, the synthetic text. Quite the contrary. It's that my verbal big ideas pieces, in the final analysis, have mostly not been quite good enough to be satisfying. Getting my hands dirty with difficult primary research helps me do them better. This is because science is a specific and concrete endeavour, and hence doing the specific and the concrete is a way of disciplining one's grasp of it. If you work on animal behaviour, then hours spent observing your animals are never wasted. If you do social research, then hours in your field site are what keep you sharp. And analysing your own data, as well as warding off dementia, brings the possibility of seeing new patterns and hence growing as a theorist as well as a data analyst. The mind fed on its own devices can become flabby and tendentious: only through a practice of repetitive confrontation with the primary phenomena we are allegedly talking about is it honed, and its confirmation biases challenged. Your animals or your people have a way of doing something you didn't expect: this is the source of a new idea or interest. Your data will be messy, and will do a better job of refining your ideas than the peer-reviewers or conference debates can ever do. And I suppose, beyond all this, doing the primary activities of your research area is simply a way of keeping busy, keeping away from too many low-quality hours spent in front of a computer. It is a way of executing Robert Burton's famous anti-depressant maxim: Be not idle![5]

Keeping your hands dirty also means learning how to do new things. And this is a good thing: the skills I picked up in graduate school could not possibly have sustained me this long. Learning new skills has always paid dividends of one kind or another; and stepping back from doing primary research myself has always been the point at which things have started to go less well.

§

I learned lessons 1-4 by through making the best of an often-bad job. These are not necessarily good ways of being a researcher, I thought, so much as good ways of managing to remain a researcher despite being as neurotic, hyperactive and easily-discouraged an individual as I am. But, reading back over them, perhaps I am doing myself a disservice. Perhaps lessons 1-4 (or lesions 1-4, as my word processor seems to want to call them) are more generally useful.

Take lesson 1, for example. Given the enormous increases in the efficiency with which we can gather and analyse scientific information in the last couple of decades, the productivity of the academy ought

---

5   Burton, R. (original publication 1621), *The Anatomy of Melancholy, What It Is: With All the Kinds, Causes, Symptomes, Prognostickes, and Several Cures of It.*

to have increased many-fold. I am not sure it has; even if the volume of output per researcher has increased, I doubt the depth has. This phenomenon is much discussed in the business literature under the name of the internet or Solow paradox: 'You can see the computer age everywhere except in the productivity statistics'.[6] For most academics, what the internet age has brought is mostly an increase in the available ways of treading water in low-quality work-related activities, without getting round to much real work. Apparently the average business email is read 6 seconds after being delivered. Given reasonable assumptions about the duration of undivided attention that useful thought requires, this means that on the modal day in a modern office (and probably university too), the amount of quality work done is...erm, none at all.[7] So if we all started our days with a few hours in which the internet was shut off and curfew was enforced, I think our outputs would increase dramatically in both quantity and quality. Interestingly, over the years I have read literally dozens of institutional plans for improvement in output. They are full of meaningless statements like: 'We will focus on our core themes whilst also responding to strategic opportunities' or 'We will expand our teaching offer whilst increasing our research capacity'. In other words, there is no strategy at all. Never once have I read one that said: we won't hold any meetings in the morning, so that staff can actually get some work done. I wonder why, since it is what might actually make some difference, and it would probably make us all nicer people to work with.

Now consider lessons 2 and 3. People straining after high impact factors and flashy publications has a serious distorting effect on scientific knowledge, and reduces the efficiency of science. It means people over-invest in under-powered, cute exploratory studies, and under-invest in well-powered confirmations. It leads to serious publication bias away from the null hypothesis, and consequent falsehood of much of what we find in the textbooks. It motivates researchers to oversell their story, and exercise degrees of freedom in what they report; and peer reviewers to focus on grandstanding and subjective value judgments, rather than providing technical verification and assistance to the

---

6   Solow, R. (1987). We'd better watch out. *New York Times Book Review* July 12, p. 36.

7   See Alter, A. (2017). *Irresistible: The Rise of Addictive Technology and the Business of Keeping Us Hooked* (London: Penguin).

authors in better understanding their data. It encourages secretiveness and competitiveness, rather than what science should be about, which is open sharing and collaboration. A bit more indifference to the glittering prizes, and more of a focus on creativity, integrity and openness, would be to our common good.[8]

And then we come to lesson 4. I recognize in myself that I have been overly tempted to perorate, at the expense of detailed empirical or computational work. Looking around, I can see a number of eminent people in my field who have made precisely the same mistake (they tend to be men, interestingly). They accept the quick and dirty from their lab or field site, or just give up on having a lab or field site at all, and carve a niche of sitting in their studies putting the discipline (or several disciplines) to rights in a stream of long-form verbal salads. Of course I see what motivates them to do this—here I am trying to do the same, after all. Look at Darwin, they say: it's the big ideas pieces that change the world. Look at Darwin, I respond. Literally thousands of hours of experiments on barnacles, worm-casts, and the germination of seeds immersed in salt water for every big ideas piece he wrote. It's the careful artisanal practice that makes the big ideas pieces really good when they come. The idea that you might only write big ideas pieces seems like an athlete choosing only to run track finals, and never training runs.

When I look at eminent colleagues who rose above the level of getting their hands dirty and became full-time commentators in the field, it seems to me that their ideas contributions started to get less valuable around the time their direct involvement in primary research reduced. Where formerly their thought was taut and rigorous, it became vaguer, flabbier, more programmatic, and more self-referential. They cherry-picked their examples. Their ability to see both sides of the problem decayed. Empirical research, I like to think, is like an adversarial collaboration with reality. The mind is like the immune system; to function properly it needs to be constantly challenged by data. So if like me you are prone to covet big ideas and the freedom to spend all day

---

8 For the critiques of current scientific practices on which this paragraph draws, see: Young, N. S., J. P.A. Ioannidis and O. Al-Ubaydli. (2008). Why current publication practices may distort science. *PLoS Medicine* 5, 1418-22, https://doi.org/10.1371/journal.pmed.0050201; and Higginson, A. D. and M. R. Munafò (2016). Current incentives for scientists lead to underpowered studies with erroneous conclusions. *PLoS Biology*, 14: e2000995, https://doi.org/10.1371/journal.pbio.2000995

pontificating, it would probably not be bad to force yourself to spend, say, two thirds of your effort gathering and analysing primary data. It is not that you shouldn't write your big ideas pieces. You should. It is that these will be improved by grappling between times with the real concrete problems of the working researcher. I like to think that Spinoza's lens-grinding did more than buy him the freedom to pursue his philosophy. I like to think it made his philosophy better.

On which note, I have spent too much time on this essay. I've got a data set that needs analysing.

14. Morale is high
(since I gave up hope)

> …there ariseth in his soul many fears, and doubts, and discouraging apprehensions, which all of them get together, and settle in this place
>
> –John Bunyan, *The Pilgrim's Progress*

> …for I feel in me
> An inexpressive lightness, and a sense
> Of freedom, as I were at length myself
> And ne'er had been before.
>
> – John Henry Newman, *The Dream of Gerontius*

In April 2015, Richard Horton wrote as follows: 'The case against science is straightforward: much of the scientific literature, perhaps half, may simply be untrue'. Horton goes on to provide a worrying charge-sheet: scientists typically leap to generalisations from overly small samples, and are abetted by the establishment in doing so; they pursue dubious trends for extra-scientific reasons; statistical inference is poor and formulaic; data are sifted to support predictions; predictions are altered to retrofit data ('These are our values, and if you don't like them…we have others!'); researchers are driven by the maximization of their own status metrics; there are often blatant conflicts of interest; universities behave like sweat shops for making more, rather than more credible, scientific outputs; scientific journals are for-profit entities that want to attract attention to their brand, not reveal the truth about the universe. The consequence is a scientific literature much of which we should be very careful about trusting.

Just what we needed, you may be thinking. Some anti-science nut, chaining together half-truths and conspiracy theories in order to undermine the case for more public investment in science, evidence-based public policy, or the teaching of evolution. We need to be out

 https://doi.org/10.11647/OBP.0155.14

there defending the enlightenment and its progeny against this kind of flat-earth knavery, which has a very definite agenda of its own. But Richard Horton is most certainly not an anti-science nut. He's the editor of *The Lancet*, one of the pre-eminent medical journals in the world. What he is reporting on in this particular editorial is a symposium involving the major funders of biomedical research, as well as some of the most senior individuals in the field, to consider 'the idea that something has gone fundamentally wrong with one of our greatest human creations'.[1] That's our real problem, you see: it's not just the barbarians outside the gates saying that the empire is decadent and corrupt. Increasingly, there is unease among the citizens inside the gates too.[2] This unease has been preoccupying me. I don't just mean that I have been reviewing my own working practices to understand how they could be more robust, though I have been doing this. I mean something deeper: it has been affecting my morale, my ability to carry on.

I attempted to go into science out of a very naïve, very pure and rather spiritual sense of love. I was always interested in the arts too, and in fact I worked in the arts for a few years. But I fell in love with science through the enchanting writings of authors such as Richard Dawkins and E. O. Wilson.[3] I can hardly describe the exhilaration of learning about science from these masters: not just that it made the drugs and

---

1   Horton, R. (2015). Offline: What is medicine's 5 sigma? *The Lancet* 385: 1380, https://doi.org/10.1016/s0140-6736(15)60696-1. By the way, the title of this essay comes from that of a show by Powder Keg theatre company, a show that was about searching for something to cling to amidst mess and uncertainty.

2   Non-exhaustive list of key references on science's current troubles: Ioannidis, J. P.A. (2005). Why most published research findings are false. *PLoS Medicine* 2: 696-701, https://doi.org/10.1371/journal.pmed.0020124; Simmons, J. P., L. D. Nelson and U. Simonsohn. (2011). False-positive psychology: Undisclosed flexibility in data collection and analysis allows presenting anything as significant. *Psychological Science* 22: 1359-66, https://doi.org/10.1177/0956797611417632; Prinz, F., T. Schlange and K. Asadullah. (2011). Believe it or not: How much can we rely on published data on potential drug targets? *Nature Reviews Drug Discovery* 10: 712, https://doi.org/10.1038/nrd3439-c1; Open Science Collaboration. (2015). Estimating the reproducibility of psychological science. *Science* 349: aac4716, https://doi.org/10.1126/science.aac4716; Smaldino, P. E. and R. McElreath. (2016). The natural selection of bad science. *Royal Society Open Science* 3: 160384, https://doi.org/10.1098/rsos.160384; Higginson, A. D. and M. R. Munafò. (2016). Current incentives for scientists lead to underpowered studies with erroneous conclusions. *PLoS Biology*, 14: e2000995, https://doi.org/10.1371/journal.pbio.2000995; Young, N. S., J. P. A. Ioannidis and O. Al-Ubaydli. (2008). Why current publication practices may distort science. *PLoS Medicine* 5, 1418-22, https://doi.org/10.1371/journal.pmed.0050201

3   For example, Dawkins, R. (1998). *Unweaving the Rainbow* (Boston: Houghton Mifflin) and Wilson, E. O. (1998) *Consilience: The Unity of Knowledge* (New York: Knopf).

the computers work, not that it added to the size of the economy, but the sweeping intellectual and even aesthetic case that underpinned it. Science: an unbounded golden web of elegant theory, beautiful experimentation, and the best of the human potential. A shibboleth that makes us scientists different from, and, frankly, better than, creationists on one side, and postmodernists on the other. The 'science'/'everything else' division became for me, I now see, the division between the sacred and the profane, remade by these great writers in a new and astonishing way. Wilson, in *Consilience,* stated very clearly that science is a qualitatively distinct kind of activity from other expressions of human belief. Other belief systems may serve 'psychological functions', he concedes, but science is revolutionary in its ability to discover truth. The Enlightenment is a singularity, and science is a new phase of human life.

This is why the current problems in science are so unsettling. To discover that the revolutionary sacred activity probably misses truth at least as often as it hits, not just through bad luck but through systematically stupid and bad behaviour; to discover that all kinds of 'psychological functions' such as confirmation bias, protection of fiefdom, the quest for status, exaggeration of a case in order to market a product, and so on, are deeply embedded in the one institution supposed to be different; what this adds up to is discovering that the ordinary, disappointing regularities of the profane are right there in the heart of the sacred. This poses the question 'How can I carry on?'. Although various theories posit income-maximisation or cultural conformity as prime movers of human behaviour, my personal experience is rather different: people, including me, want to do things that they could readily justify to a jury of their peers (including, critically, the jury within). The personal cost of trying to do science rather than something else is very big. You have to feel convinced it is worth it. You need to know that the things you believe and promote have some validity. You need to feel sure that it isn't all some kind of delusion, quackery, or racket.

I have two questions today. The first is really the warm-up: how can we simultaneously accept the evidence that the actual practices of science are flawed, and its products often wrong; and yet hang on to the assurance that science is a special kind of activity whose long-term arc bends towards the truth? The second is the small matter of how, having answered the first question, we can best live.

§

The first question turns out to be surprisingly easy. In order for the long-term arc of science to bend toward the truth, science does not have to be perfect. It only has generate a force that is on average very slightly stronger than the forces that hold human knowledge back. That's how science can be both very similar to other kinds of human activities (shamanism, rhetoric, marketing and what not), and also revolutionarily different. Those other activities all generate a velocity slightly less than that needed for epistemic escape; science generates a velocity that is at least sometimes slightly greater. A small difference, with big consequences.

Let us spell this out with an example: powered flight. What was revolutionary about early powered aircraft is not that they were efficient. They weren't. In fact, they were terrible. The vast majority of the energy they generated was wasted as heat. Of the energy they did manage to generate as motive force, very little was converted into lift. So the point was not that they were very good. They were about the worst devices for powered flight you could come up with, except for all the other devices that had been tried out in the history of humanity. For most of those earlier devices, the lift they produced was insufficient to exceed the pull of gravity. The early powered aircraft were only incrementally different, perhaps, but the increment was a consequential one: it was the increment that reversed the sign of difference between gravity and lift, not by much, and not always, but enough for something unprecedented to happen. And once the sign was reversed, once the planes took flight, their design could be gradually improved by the cumulative tinkering that characterises human culture.

Let us return to science. It is not that the people, or even the institutions, that characterise science as a profession are so very different from any other body of people or institutions. It's that somehow, the interaction of those people with those institutions has led to a slow accretion of better understanding of the world over long passages of time. Much of science's energy is wasted: the ideas and the claims in any individual publication or even career mostly turn out to be nugatory. But the resultant of all the chaotic motion is a ratchet of gradually better understanding of the processes of the world. The good stuff is just slightly more likely than the bad stuff to be generated and retained, on average. The improving arc is more perceptible the further away you stand: close up, you only see the individual sparks flying off in all directions, mostly not the right one. Only from afar do you see that there is a bit more energy going in one direction than in the others. Just as in the powered flight example, once a science has achieved some kind of lift off, its efficiency can be cumulatively improved. We should be putting as much energy into reforming methods (improving the efficiency of plane design) as we do into individual studies (going for a fly around). The current debates within the scientific community, the so-called 'replication crisis', should really be seen as discussions about how best to do this, not repudiations of the whole scientific enterprise.

This view of science leads very naturally to seeing science-ness as a continuum. The best cases for the revolutionary nature of science can be made from physics, from chemistry, and from certain parts of basic biology. It's no accident that the best cases made by Dawkins and Wilson come from those areas of science. The day-to-day reality of my working life, though, comes from the study of behaviour and society, where the situation is rather less decisive. If lift routinely and decisively exceeds gravity in physics and chemistry, then the two forces are much more nearly equal in the social and behavioural sciences. Where the two forces are about equal, there is a lot of scope of bump along with bad ideas persisting too long; multiple incompatible views being held simultaneously; fads that appear and vanish like the morning mist; and rhetoric, ideology, and social influence determining the disposition of the field. That's why professional disputes are often so prolonged and so bitter in social and behavioural fields: because, as it were, the stakes are so low. Still, we have to hold on to the hope that even in these fields, the arc towards the truth proves a bit stronger than the will to power in the very long term.

§

The problem of how to live, as a scientist, is the following one. You need faith in order to be able to do the work. Faith that what you are doing is sensible and worthwhile. Faith that you have the right methods and design. Faith that the patterns you see could be real patterns. Faith that the way you have analysed your data is a sensible way. Faith that the arguments you make are good arguments, and important ones. You need faith in all these things because the whole process is genuinely difficult, and very slow; you are constantly knocked off course by obstacles and distractions; peer reviewers can be quite gratuitously unpleasant, as well as sloppy; rejection is designed in to the process; and employment conditions are often less than ideal. So if you do not have sufficient faith in what you are doing, you will quite sensibly walk away.

Yet science is a system of organized scepticism. Faith is the one thing you should never have. The view of science I have sketched in the last few pages suggests that, rationally, you should make the pessimistic meta-induction: the specifics of the thing you are working on will probably not turn out to be as you believe them to be; your results will probably not replicate; your methods will appear naïve and flawed to a future generation; and the world will not turn out to be quite as you contend. It's a hard thing to pull off this trick: enough conviction to get up and go to work every morning, enough scepticism to remain a good scientist.

The way most successful scientists solve this is very simple: they have faith in their own work, and relentless scepticism about everyone else's.[4] It's much like the large majority of car drivers who apparently believe their driving to be better than average. But we are all grown-ups here: we should know that we can't all be better than average. The lesson I take from the replication crisis is not that there are few bad apples in science who should be pilloried. It's that we've *all* been doing bad science, probably still are, in myriad banal ways that are so habitual that we don't even realise their significance. The very fact we can't see anything wrong with our own practices is precisely the point. Though we understand that *others* might fudge the theory, hypothesise after the results are known, exercise researcher degrees of freedom, or torture the analysis in search of the 'significant' *p*-value, it doesn't feel like we *ourselves* do it. But we must concede there are thousands of tiny judgements involved in the writing of every single paper, the analysis of every dataset, and these are not generally recorded in any public ledger. And the thing about self-deception is that you are always the last person to know.

The prospect of patterns in my empirical work not replicating holds a particular awfulness for me, a feeling of bleeding slowly to death. And just as I ought to expect, there have been some instances of it. Let me give you an ongoing example. Melissa Bateson, myself, and colleagues had

---

4   This brings us back to the essay with which this book began: *How my theory explains everything: And can make you happier, healthier and wealthier*, this volume.

a beautiful finding in starlings. We'd been measuring a physiological marker that relates to future life expectancy. (The marker is called developmental telomere shortening, and it measures how much the telomeres—the DNA caps on the ends of chromosomes—of red blood cells shorten over the bird's early life. More shortening predicts shorter subsequent life.) We then conducted a behavioural experiment in which we assessed how impulsive the birds were; how strongly they preferred a small but immediate reward to a larger reward for which they had to wait longer. It's basically the famous 'marshmallow test', but for starlings.

In our first experiment, we found that birds with more developmental telomere shortening were also more impulsive. But of course, that makes perfect sense: if you are going to live longer, you can afford to wait, whereas if your life-span is limited, you'd better take what you can get now. It is compelling; somehow the birds can detect their own somatic state, and set their behavioural priorities accordingly. This beautiful result became a much-cited publication.

We have repeated the experiment twice more, in different sets of birds.[5] Figure 6 shows the results in the form of what is known as a cumulative meta-analysis. To orient you, the first row on the figure shows the result of the first experiment, described as '2012' because that is the year the birds hatched. The square indicates that there was an association between developmental telomere shortening and impulsivity in that experiment with a regression coefficient of around 0.5 (zero means no association; the larger the coefficient, the stronger the positive association). The horizontal whiskers show the 95% confidence interval for that coefficient; the zone in which, in light of the experiment's findings, we should believe that the 'true' association between developmental telomere shortening and impulsivity in starlings falls.

---

5   The sources for this section are as follows. 2012 birds: Bateson, M. et al. (2015). Developmental telomere attrition predicts impulsive decision-making in adult starlings. *Proceedings of the Royal Society B: Biological Sciences* 282: 20142140, https://doi.org/10.1098/rspb.2014.2140; 2013 birds: Nettle, D. et al. (2015). Developmental and familial predictors of adult cognitive traits in the European starling. *Animal Behaviour* 107: 239–48, https://doi.org/10.1016/j.anbehav.2015.07.002; 2014 birds: Dunn, J. et al. (2018). Early-life begging effort and adult body condition affect choice impulsivity in the European starling (Sturnus vulgaris). Unpublished manuscript: Newcastle University.

Our first follow-up experiment was on another cohort of birds hatched in 2013. The second row on the figure shows the association and its 95% confidence interval when the data from the 2012 and 2013 birds are combined. We then performed another follow-up on a cohort of birds hatched in 2014. The third row adds the data from these birds in to the other two cohorts. I think you can see what is happening. As we gather more data, the apparent strength of the association is bleeding away toward zero. (For God's sake, don't anyone do a fourth experiment!) The pattern shown in figure 6—the first-published associations are the strongest, and as more evidence comes in, the overall association gets weaker and weaker—is frequently found in scientific literatures. It is attributed to differential likelihood of flukily strong early results being published and getting notice, followed by the gradual dilution of these early outliers by more typical replications. But to achieve it within your own lab, well that's quite something.

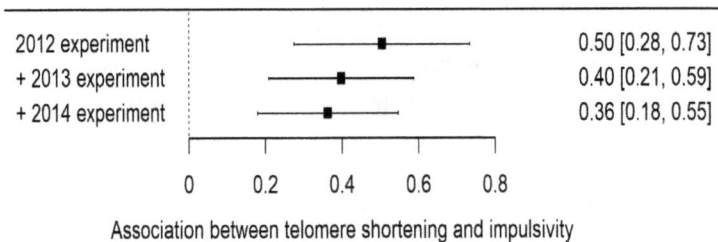

Figure 6. Forest plot from a cumulative meta-analysis of three successive experiments on developmental telomere shortening and impulsivity in starlings. The squares and whiskers indicate the regression coefficient of the association between developmental telomere shortening and impulsivity, and its 95% confidence interval. These are also shown as numbers on the right. The first row represents the experiment on the 2012 birds alone; the second row the 2012 and 2013 birds combined; and the third row the 2012, 2013 and 2014 birds combined.

Rationally, I don't think we have anything to reproach ourselves for. We asked a fair question, and Nature is answering in her own sweet time. There might still be something to our initial finding. Actually, even though the impulsivity results have not held up as we hoped, we have found evidence from other, different behavioural tasks suggesting that birds' behaviour may be related to their future life expectancy, as measured by developmental telomere shortening.[6] The follow-up impulsivity experiments are not exact replications of the original (the developmental histories of the birds were rather different in the three cohorts, and in 2014 the impulsivity measure was different too). Perhaps we have hit upon some interesting and real heterogeneity. And we have not tried to suppress or massage away the inconsistency of the results: we should perhaps get some credit for this. But still: it's devastating, and it keeps me awake at night. Why?

Part of the reason is to do with the usual human concerns: self-consciousness about reputation, status and apparent competence. Feeling like no-one will ever believe anything I say again. I am the first scientist in my family (when a teacher suggested at a parent-teacher evening that I should consider further study of mathematics, my father queried whether they were discussing the correct child). I have never felt quite at ease in the august and self-confident circles in which I can nowadays move, so there is more than a hint of the imposter's fear of being found out. I think of my dream-friend Franz Kafka. The key to understanding Joseph K.'s odd passivity when his accusers arrive at his door in *The Trial* is that somewhere in his heart, he already suspected he was guilty. In some diffuse way, he was expecting it. He just didn't yet know what it was he was guilty of.[7]

But there is more to it than this. When you analyse a dataset, there is a lot of tedious merging and cleaning and preliminary analysis. Then at a certain point, you try an analysis or two, and suddenly see a pattern. That moment is psychologically completely and utterly compelling. The pattern jumps out at you with a concrete and immanent reality. You instantly grasp why that is the pattern that makes sense, that made sense all along. This is an example of what Nick Chater calls the 'grand illusion' of consciousness: when our minds alight on a belief or percept, it feels as if we had always had that belief or percept there, waiting in

---

6   Nettle, D. et al. (2015). Developmental and familial predictors of adult cognitive traits in the European starling. *Animal Behaviour* 107: 239–48, https://doi.org/10.1016/j. anbehav.2015.07.002; Andrews, C. et al. (2018). A marker of biological ageing predicts adult risk preference in European starlings, Sturnus vulgaris. *Behavioral Ecology* 29: 589–97, https://doi.org/10.1093/beheco/ary009

7   He still didn't know as he was being repeatedly stabbed to death in a quarry. This is possibly an analogy I should not extend too far.

the mental depths to be brought into the foreground. No matter that, in reality, the brain may have fabricated it that very instant.[8]

It's very hard to tell yourself, in the grip of the grand illusion, that the result may be a fluke; that you may have tortured the dataset until it confessed to something; that your prediction has shifted and you are rationalizing yourself after the fact. When someone else's subsequent experiment, or your own, fails to reproduce the finding, it is as unnerving as the discovery that a conversation you had yesterday was in fact a hallucination. But I saw it! The *p*-value was there! Oh please, let me not be mad!

§

When middle-aged scientists go through replication crises, some turn to drink, some to drugs, some to Bayesian statistics. I have been turning to Buddhism. I can't say I have got very far (aren't there a lot of numbered lists?). Nonetheless, there are some elements of Buddhist thought that seem like they could come in handy in getting papers written and equanimity preserved.

At the heart of Buddhist thought, as far as I understand it, lies a network of linked ideas. The first, expressed in the first noble truth, is that living is synonymous with suffering, or at least, exposure to the possibility of suffering (*dukkha*). This is because, in living, we crave and are attached to—indeed are fuelled by—worldly things that are by their nature impermanent and likely to fail us or fade away. These worldly things include pleasure, material goods, status and renown, but also beliefs and habits. To the extent that we condition our happiness on obtaining or maintaining these things, we are locked into a cycle of endless living-suffering (*samsara*), because pleasures always fade, beliefs turn out to be wrong (see figure 6), and status is never enough.

This is the bad news, but the good news follows: once we recognise the reality of suffering, and its causes, we see that it can also cease, and that there is an available route to liberating ourselves from it. This route requires nothing other than enlightenment of our minds. We achieve the liberation not by satisfying our attachments and cravings, which would after all just bind us further into *samsara*, but by living according

---

8   Chater, N. (2018). *The Mind is Flat: The Illusion of Mental Depth and the Improvised Mind* (London: Penguin).

to the eight-fold (or just possibly three-fold, or 8 x 3 = 24-fold) noble path.[9] This path consists of right speech, right action, right livelihood, right effort, right mindfulness, right concentration, right view, and right resolve. Basically, trying to live mindfully, wisely and well. The noble path leads directly to the release from the compulsory cycle of desire, striving, attachment and suffering known as *nirvana* (often translated in US English as 'tenure').

The noble path is not just asceticism. The Buddha certainly rejected living by trying to fulfil one's cravings as unprofitable and painful; but he rejected a life of extreme austerity and self-mortification too, on the same grounds. After all, attachment to austere ritual, to self-denial, is just another form of attachment. Instead he proposed the middle way: in the world, of the world, yet trying to navigate it nobly.

Before I completely lose any credibility I had, I would invite you to consider the above doctrine not as supernatural or even religious, but as a set of rules of thumb for living worked out over centuries by thoughtful members of a smart species of ape.[10] A species with no single mental governor, but whose mind consists of a noisy parliament of different and perfectly explicable motivations—for resources, for reputation, for sex, for power, for avoiding danger—which together conspire to produce powerful patterns of habit and thought, patterns that can in the long term produce interpersonal and intrapersonal problems. But this same species of ape also, for extraordinary but not supernatural reasons, possesses a surprising capacity for offline reflection and reasoning, a capacity that can be used to calm fractious disputes and reach wise compromises, including, critically, disputes and compromises within the parliament of the mind. The first noble truth reminds us that our very real and natural motivations can make us disappointed or miserable and cause us problems; the noble path reminds us that we have mental resources to deal with these problems, and it's a good idea to practice using them.

Let's apply some of these ideas back to the life of the scientist. To be a scientist is synonymous with suffering, or at least exposure to

---

9   See Hanh, T. N. (1998). *The Heart of Buddha's Teaching: Transforming Suffering into Peace, Joy and Liberation* (New York: Random House).

10 This is the approach of Jonathan Haidt's 2006 book *The Happiness Hypothesis: Finding Truth in Ancient Wisdom* (New York: Basic Books).

the possibility of suffering. The reasons for this are part banal—we all want the big grant, the big promotion, our papers to be accepted in the selective journals, but we can't all achieve this. Others will succeed and mostly we will fail. There are more metaphysical reasons, too: the beliefs and hypotheses to which we devote thousands of difficult hours of our lives will more often than not turn out to be wrong. We won't know this for a very long time, perhaps never, and, individually, will have very little control over how it turns out. So we must live with doubt and uncertainty about the aspects of our lives that are more important to us than anything else. And what we achieve always, in my experience, falls short of what we hoped to achieve: as E. O. Wilson put it, all scientists 'are children of Tantalus, frustrated by the failure to grasp that which seems within reach'.[11] Thus, if we base our well-being on getting what we crave, or attachment to what we have done before, we can never be really satisfied, for we are trying to hold on to a will o'the wisp.

Faced with this dilemma, two courses suggest themselves. The first is equivalent to the life of hedonic gratification: in the long run, no-one knows who is going to be right, and I won't be around to find out anyway, so I will just make as good a career for myself as I can. Consider researchers of type A. They make a big name for themselves with their seminal *Hypothetical Attachment Theory (HAT)*, or whatever. They make the positive case for *HAT* in big idea piece after big idea piece. They churn out empirical studies, and present them in the best pro-*HAT* light the peer reviewers will let them get away with. The studies are as good as they need to be, but no better. They aggressively confront journal editors who reject their papers. When peer-reviewing, they recommend 'reject' if the authors do not cite enough *HAT* references, and that is nearly always. They know who their rivals are and make sure to rubbish their grants so that they do not get funded. As the evidence accumulates that *HAT* is probably old hat, they dismiss the criticisms as ill-founded or personally motivated. They defend their dung-hills. They see the exchange of academic views as a social game to be won. It is because senior people are type-A researchers that science sometimes seems to advance, in the famous paraphrase of Max Planck, one funeral at a time. It is obvious that type-A researchers suffer from grasping attachment:

---

11   Wilson, E. O. (1998). *Consilience: The Unity of Knowledge* (New York: Knopf, p. 3).

to status, power and their habits of thought. They are locked into the *samsara* of revise and resubmit, busy-work, dogma and self-promotion.

Now consider researchers of type B. All too aware of the possibility of error and luck, they are desperately cautious about what they will publish. They always want to take more time over everything. They need more data, another replication, before they will show their results to the world. They always need to do more reading, in case there is something they have missed. And when they do write, there are caveats on their caveats: their papers will not come down strongly for any position, all too aware that to do so might lead to saying something that could turn out to be wrong. The type-B approach is also problematic. Science could not work if everyone followed it. The dynamism of science emerges from it being a vigorous and passionate conversation, with people willingly sticking their heads above the parapet with interesting data and the strongest possible advocacy of particular ideas. And, less obviously, researchers of type B are suffering from grasping attachment too, like the ascetics criticized by the Buddha. In fact, it's another form of egotism. They crave a kind of certainty and definitiveness that we can never really have; they are too attached to their own self-image, their personal comfort and their rituals of scholarship to be prepared to let the ideas and the data speak for themselves.

You can see where I am going here: towards the middle way, and a path to the cessation of suffering. The middle way is where we are quite prepared to put out what we have done, including strong and principled, even passionate, arguments for what we think it means theoretically; yet on the other hand we are open to changing our minds at a moment's notice; we encourage alternative views and welcome those whose starting point is different from our own; we are quite prepared to say when we were wrong, and patient to say why if we still think we might be right. Really it comes down to humility and openness: Open sharing of our data, openness about what operations and analyses we have performed, openness to sticking a preprint out there that turns out to be ill-informed, openness to alternative views, openness to trying to see things a different way. The eight-fold noble path (right speech, right action, right livelihood, right effort, right mindfulness, right concentration, right view, and right resolve) is a pretty good recipe for making contemporary science not just more effective, but also a nicer place to live. We need to provide both the incentives and the social norms that encourage this kind of nobility, and discourage its opposite.

§

What, then, of *nirvana*? This is often translated as 'release from the cycle of existence'. I have always struggled a little with this. Many of my very favourite things depend rather heavily on me existing. So why would I follow a path whose best-case scenario is self-obliteration? This question, in the general case, is rather above my eschatological pay-grade, but we can make a useful translocation of the *nirvana* concept for science. In fact, there is a surprising link to Karl Popper, of all people.

In contemplating the human capacity for reasoning, and hence science, Popper made the following point (and in doing so revealed himself as a better philosopher than zoologist). When most animals hold a false belief, it can lead to their deaths (think of a deer with a false belief about which species are dangerous predators, or a false belief about which food is safe to eat). So the stock of beliefs is only improved by the cycle of birth (which introduces variation in beliefs), and death (which disposes of the false ones). The miracle of being human rather than being some other kind of animal is that the deaths of our ideas can become decoupled from our own deaths. We can represent ideas symbolically, then debate, converse, test, adjudicate, modify, falsify, and eventually reject them, all in relative safety. We can, as Popper put it, 'let our false theories die in our stead.'[12] This opens up the possibility of an adaptive evolution of ideas, with a generation time much faster than our biological generations, but giving ideas an eventual lifetime that could be much longer than our own. Ideas take on a life of their own. Science, perhaps above all else, is the commitment to fostering this artificial life: ideas proliferating, mutating, recombining, dying and becoming immortal in the rich, distributed ecology of the scientific literature.

What does this have to do with *nirvana*? The following: what is it that a scientist can most sincerely hope for? What most can sustain him in feeling that it was all worthwhile? It is not being right. It is not being the cleverest. It is that the ideas to which he devoted his life ultimately

---

12   Magee, B. (2010). *Popper* (London: Fontana, p. 64).

released him from their cycle of birth and death. Those ideas began to evolve on their own, in ways he could not predict, through other people, through artefacts and dusty books, through conversations in coffee shops, through different technologies or social institutions, through a shy student's first moment of enchantment. This possibility is profoundly and existentially comforting though, paradoxically, the person for whom it is comforting vanishes from the picture.

This is what I think. It's a long old game. If I stick at it well enough, I will cease to matter. I will sooner or later disappear to the hills, another wind-lined Pennine man; inconclusive; increasingly vague; stalwart of the local choral society; heft gradually diminishing through ordinary somatic processes. But maybe I'll know that, somewhere in the world, these ideas that I have cherished, these ideas will be dying—and living—in my stead.

Acknowledgements

Hanging on to the Edges has been a long journey. I would like to thank Andrew Scott, in whose company the idea for the project was born. Thanks, too, to our lively research group at Newcastle University, for their hard work, good company, and conversation; this includes the generations of students and interns who have passed through. The European Commission has supported our lab financially through its Horizon 2020 programme, and so in an indirect way made this book possible as well. My lovely collaborators and colleagues are too numerous all to be named here, but I would like to single out Willem Frankenhuis, for discussions on many of these topics; Gillian Pepper, for her work that underlies The mill that grinds young people old in particular; and Graham Crow, for his inspiring enthusiasm about building bridges.

Many people provided comments on individual essays when they were first published online. These were a great source of encouragement and improvement. Hugo Mercier, Nichola Raihani, Rebecca Saxe and David Lawson read the full draft of the book version. I have not always followed their advice, and all shortcomings and idiocies are of course my own.

I'd like to thank everyone at Open Book Publishers for their remarkable enthusiasm and efficiency, and for making books available for free to students and scholars all over the world. I am particularly impressed that my editor picked up the fact that I had got the gender of the villain wrong in an allusion to a fairly obscure 2007 horror movie.

Pat Bateson died half way through the writing of *Hanging on to the Edges*. Some of it might have puzzled him: his pre-replication-crisis generation was less afflicted than mine is with doubt and uncertainty about their science (or maybe that's just how Cambridge people are). But much of it—on the active role of the individual organism, on the need to transcend the boundaries between the 'biological' and the 'social', on the aspiration of science to do good in the world—is a testament to what he believed in, and patiently argued for, over many years. He influenced me both directly and indirectly.

Finally, my greatest thanks, and my love, go to Melissa Bateson, my inspiration, principal scientific collaborator, spouse, and fellow adventurer.

Index

This book need not end here...

Share

All our books—including the one you have just read—are free to access online so that students, researchers and members of the public who can't afford a printed edition will have access to the same ideas. This title will be accessed online by hundreds of readers each month across the globe: why not share the link so that someone you know is one of them?
This book and additional content is available at:
https://www.openbookpublishers.com/product/842

Customise

Personalise your copy of this book or design new books using OBP and third-party material. Take chapters or whole books from our published list and make a special edition, a new anthology or an illuminating coursepack. Each customised edition will be produced as a paperback and a downloadable PDF.
Find out more at:
https://www.openbookpublishers.com/section/59/1

Like Open Book Publishers

Follow @OpenBookPublish

Read more at the Open Book Publishers **BLOG**

You may also be interested in:

Tyneside Neighbourhoods
Deprivation, Social Life and Social Behaviour in One British City

By Daniel Nettle

https://www.openbookpublishers.com/product/398

Wellbeing, Freedom and Social Justice
The Capability Approach Re-Examined

By Ingrid Robeyns

https://www.openbookpublishers.com/product/682

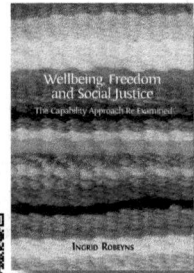

Just Managing?
What it Means for the Families of Austerity Britain

By Mark O'Brien and Paul Kyprianou

https://www.openbookpublishers.com/product/591

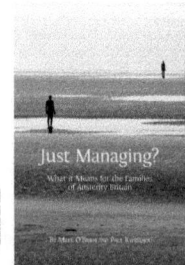

www.ingramcontent.com/pod-product-compliance
Lightning Source LLC
Chambersburg PA
CBHW071102280326
41928CB00051B/2698